Electromagnetic Phenomena and Health – A Continuing Controversy?

IOP Conference Series: Earth and Environmental Science
Volume 10

London, UK.
10 September 2008

ISBN: 978-1-5108-0201-8
ISSN: 1755-1307

Printed from e-media with permission by:

Curran Associates, Inc.
57 Morehouse Lane
Red Hook, NY 12571

Some format issues inherent in the e-media version may also appear in this print version.

Printed by Curran Associates, Inc. (2015)

For permission requests, please contact the Institute of Physics
at the address below.

Institute of Physics
Dirac House, Temple Back
Bristol BS1 6BE UK

Phone: 44 1 17 929 7481
Fax: 44 1 17 920 0979

techtracking@iop.org

Additional copies of this publication are available from:

Curran Associates, Inc.
57 Morehouse Lane
Red Hook, NY 12571 USA
Phone: 845-758-0400
Fax: 845-758-2634
Email: curran@proceedings.com
Web: www.proceedings.com

Table of contents

Volume 10

Electromagnetic Phenomena and Health – A Continuing Controversy?
10 September 2008, The Institute of Physics, London, UK

Accepted papers received: 22 April 2010
Published online: 25 May 2010

Papers

IOP Conf. Series: Earth and Environmental Science **10** (2010) 012001 doi:10.1088/1755-1315/10/1/012001

Science and Precautionary Measures in EMF Policy

Mike Repacholi

Electronic Engineering Department, University of Rome "La Sapienza", Rome, Italy.
Email: repacholi@die.uniroma1.it

Abstract. Electromagnetic fields (EMF) are generated by many of the technologies we use in everyday life. Since virtually everyone on the planet is subjected to these fields, safety from exposure to them is a matter for government health policy. Most countries limiting exposure to EMF fields have adopted those recommended by the World Health Organization (WHO); the guidelines on exposure limits published by the International Commission on Non-Ionizing Radiation Protection (ICNIRP). Yet there are challenges to the use of established science and some call for the incorporation of additional precautionary measures into government policy. This paper discusses the use of science and precautionary measures for policies that protect the health of everyone from exposure to manmade EMF.

1. Introduction

Concern about possible health effects from exposure to electromagnetic fields (EMF) have continued for many decades, ever since radars were introduced during the 2nd world war. The level of concern has oscillated over time depending on the technology of the day: from the high frequency radiofrequency (RF) fields emitted by radars and radio and TV broadcasting to microwave ovens and then to the extremely low frequency (ELF) (mainly 50 and 60 Hz) electric and magnetic fields emitted by the generation, distribution and use of electricity, and more recently to the fields emitted by wireless communications systems.

There are many people who have genuine fears about consequences to health from the EMFs emitted by these technologies. The fact that science is not able to give unconditional assurances about safety from EMF emissions has lead to many individuals calling for the use of precautionary approaches to keep emissions as low as can be achieved.

This paper is based on a presentation given at a one-day conference hosted at the Institute of Physics in London, UK. It describes the challenges to and issues for national authorities to obtain sound scientific information from which policies can be developed to protect their citizens at a time when the rapid advances in RF technologies show no signs of abating. This paper puts forward the view that established, good quality science must be a priority as the underpinning of public health policies and not the incorporation of unjustifiably stringent precautionary measures merely because some do not agree with the scientific findings. Examples come mainly from the area of telecommunications where RF emissions are the main current concern, though concerns also exist for lower frequencies of the electromagnetic spectrum.

2. Public Concerns

Over 6000 papers have been published, of varying quality, that provide information relating to biological effects from exposure to EMF. When one considers publications on any physical, chemical or biological agent, the total number of papers on EMF is second only to the number published on ionizing radiations, and has exceeded all papers on chemicals. Yet with all this knowledge and the solid scientific database that has developed from the hundreds of millions of dollars spent on EMF research, public fears continue. Some of the main reasons for this concern are:

- fear about the introduction of unfamiliar or new EMF technologies into their working or living environment, especially those technologies that, for their intended purpose, must expose everyone to RF fields to allow mobile communications to function;
- location of EMF sources close to homes or where children spend many hours each day;
- the inability of science to give definitive answers about safety from exposure to the emissions from EMF technologies;
- fear for the safety of young children who will be exposed to EMF over a large part of their lifetime, and;
- claims of detrimental health effects (hypersensitivity reactions) from various EMF sources;
- information from the media on low level effects from EMF exposures.

There are particular worries that young children, who are still growing and developing, may be more sensitive to EMF than adults. These concerns have been enhanced by the fact that science has not yet completed all the studies necessary to fully address this issue. Usually, by the time scientific studies have been completed, new technologies are marketed that may produce different EMF exposure patterns in the population or present untested individual exposure situations. In effect, a major concern has been that the technological revolution is occurring at such a speed that science seems unable to keep up; hence the need for ongoing research to ensure that the right path is being taken and the correct principles adhered to.

While some people have genuine concerns about EMF health risks, there are activist groups who misrepresent the science and create even greater fears among the general population, and others who tend to dismiss such matters out of hand instead of dispassionately examining the facts. Also EMF has become a popular media issue. It is well known that the media is at its most "entertaining" when issues are hotly debated between opposing sides, where truth is the victim and unnecessary fear and sensationalism is the outcome. While not always the case, whenever a new scientific study is published that purports to find an adverse effect from EMF exposure it normally receives strong media coverage, as opposed to a study not finding any effect at all. As the media is the main source of information about EMF, the public is often subjected to a seesaw of views about what the established science is really concluding. Unfortunately, this misinformation can have a cumulative effect, which increases the concern and anxiety.

3. Scientific Principles

When a scientific study is published with results that challenge those of a number of other studies, and needs to have a unique mechanism to explain the findings, one needs to be cautious and see if such results comply with the basic principles of science. As shown schematically in figure 1 below, physics forms the basis of chemistry, which in turn forms the basis of biology, which finally forms the basis of medicine. Hence, even though there is an increase in complexity as you move up this progression, each successive layer of the science must obey the fundamental laws found to be valid for the layer below; and the mechanisms of action should also logically fit into this pattern.

At the most fundamental level are the laws of physics, which have been exhaustively validated by experiment and through internal consistency. The principles behind RF waves, namely Maxwell's laws of electro-magnetism, are accepted to be invariant in time and space, and their accuracy in describing the interactions between EMF and matter underlies the functioning of virtually all technology. No exceptions have been found, only refinements, in spite of constant challenges. Likewise, physics has been found to be valid in complex systems, encompassing chemistry, biology, technology, and medicine. Simple conservation laws (*e.g.*, energy, motion, charge, momentum) are universally applicable, and biology is no exception [1,2].

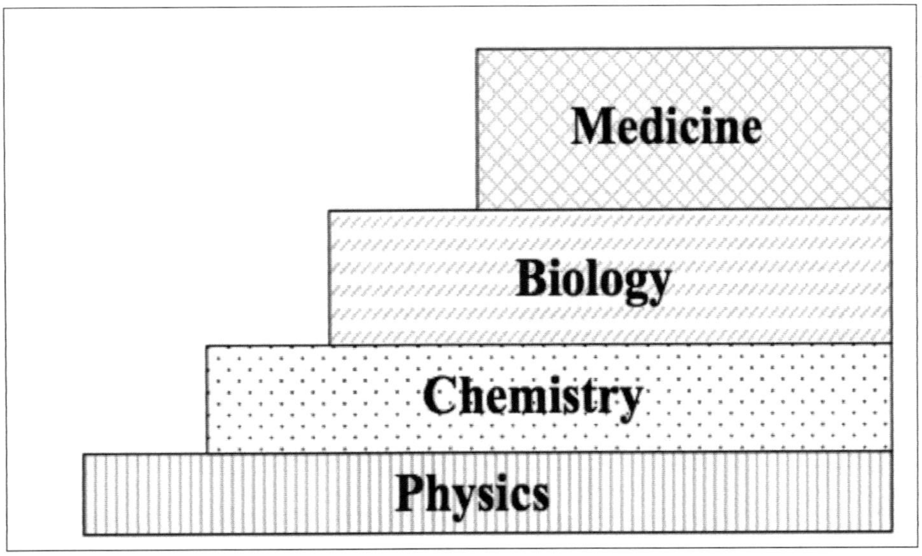

Figure 1. Each scientific discipline rests on the underlying laws of a more basic discipline (From [2])

3.1 RF mechanisms and modulation

Mobile phone signals transmit information that is encoded into electromagnetic waves by means of "modulation", which refers to the patterns of change in the frequency and/or amplitude of the RF carrier wave. As mobile phone technology has advanced, the modulation patterns have become increasingly complex, and the question has been raised as to whether a high-frequency modulated RF wave might have greater health effects than a pure sinusoidal RF wave. The applicability of fundamental physics to all systems, and to biology in particular, permits one to conclude that modulation is unlikely to give rise to unexpected RF interactions with ions, molecules, cells, and organisms, certainly at the levels of RF exposure that occur in our living environment [1-4].

Both experimental evidence and theory suggest possible demodulation of EMF waves at membrane barriers for frequencies <10 MHz, but not at higher frequencies. Although signal levels somewhat below system noise can be detected, signal-to-noise ratios substantially less than 0.1 cannot be overcome by cooperativity, signal averaging, coherent detection, or by nonlinear dynamical systems [5]. Sheppard and coworkers also concluded from their very detailed review of mechanisms that for frequencies below approximately 150 MHz, shifts in the rate of chemical reactions can be mediated by radical pairs and, at all RF frequencies, dielectric and resistive heating can raise temperature and increase the entropy of the affected biological system [5]. Thus it appears that, at the frequencies and intensities used by telecommunications systems, levels well below the international exposure limits [6], no non-thermal mechanisms could plausibly cause adverse health consequences.

Viewing modulation effects in a different way, for any mechanism capable of detecting the difference between a modulated and a non-modulated RF signal there are two possibilities: either the mechanism is fast enough to respond to and detect changes in the central RF frequency, or it is sensitive to RF power changes occurring at the modulation frequency. For the first case, scientists have not been able to identify biological structures capable of the necessary high-frequency RF tuning or bandwidth discrimination. For the second case, being sensitive to the power changes in the signal would require a "rectifying" biological structure that is non-linear at low power levels, which has not been identified by biologists or anatomists [1, 2]. Even if such a "rectification" occurred, it remains difficult to envision how a small RF power modulation input, occurring at levels below international RF exposure limits [6], could lead to any deleterious effect on biological systems, which have considerable thermal output and thermal inertia. Non-thermal mechanisms relying on non-linear responses (*e.g.*, breakdown of the cell membrane) that require RF-electric-field thresholds that are quite high and would not be expected, aside from elevated RF levels that could produce hazardous tissue heating [2].

3.2 Causal chain to produce health effects

For RF energy to change physiological function, initiate dysfunction, or cause the onset of disease in human beings or animals, there must exist a mechanism by which EMF alters molecules, chemical reactions, cell membranes, or biological structures. RF is a physical, not chemical, agent, so biological plausibility for causing health effects must be assessed with this in mind. The initial physical step is illustrated in the following causal chain by which RF interaction effects could occur:

RF⇒Matter (physics)⇒Molecules (chemistry)⇒Organisms (biology)⇒Disease

This process is illustrated in more detail in Figure 2. Biological processes in the body include many interactions among electric charges on ions, molecules, proteins, and membranes. Hence, one can imagine that exposure to RF energy that can exert forces on fixed and moving charges may have the potential to modulate biological function. For RF fields to cause or exacerbate disease in humans, RF electric fields would have to trigger not only an initial transduction step, but also begin a cascade of sequential steps that leads to a disease outcome. The causal chain, shown in Figure 2, begins with human exposure to RF fields. To complete the first step, RF fields must interact with biological molecules (or structures) in such a way as to alter their size, shape, charge, chemical state, or energy. In this energy "transduction" step, some absorption of RF energy must occur or there can be no effect [2].

For observable biological, and possibly adverse health effects to follow transduction, sequential events at the molecular, cellular, and tissue level are required, leading to a variety of possible endpoints. The outcome "Progress toward disease" in the upper right hand corner of Figure 2 is only one of many possible outcomes but it is an outcome that requires specific triggering of intermediate steps. Figure 2 illustrates multiple points in the causal chain where the signal produced by the preceding step might be within normal variations and would have no further functional consequences beyond this point in the causal chain.

At non-thermal levels of RF the first mechanistic or transduction step in the multi-step pathway has not been found to occur, aside from the well-known "hearing effect" for pulsed high-level microwaves, which has been shown to be due to the RF pulse induced thermoelastic pressure [7, 8]. Since the viability of the human body depends on the normal structure and function of large molecules (*e.g.*, proteins, nucleic acids, carbohydrates, and lipids), any proposed mechanism must predict how RF fields could interfere with or modify the normal synthesis, function or degradation of these molecules. The RF interaction mechanism(s) would then predict thresholds of exposure effectiveness in terms amplitudes, frequencies, time of onset, intermittency of exposure, homogeneity/ heterogeneity of amplitudes and frequencies, exposure duration, transients, polarization, *etc* [2].

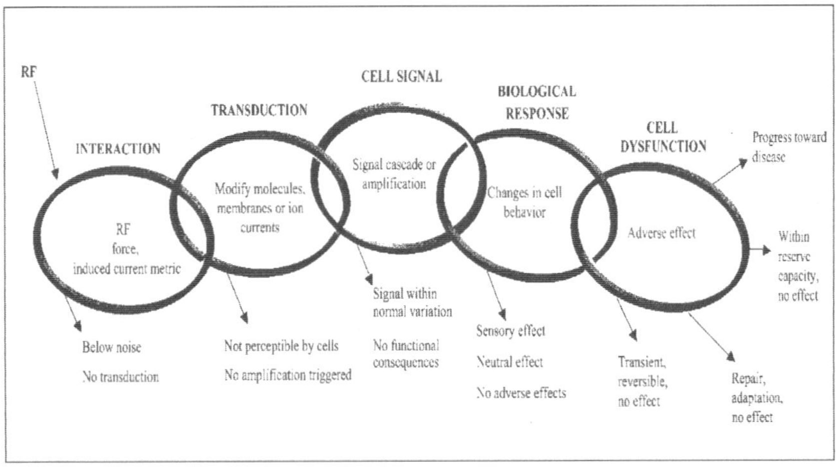

Figure 2. A transduction mechanism is only the first link in the causal chain (From [2])

Examination of the magnitude of the possible RF biophysical interactions (thermal, photon, force) shows that, under modulated RF exposure conditions allowed by the international exposure limits [6], there does not appear to be any hazard specific to RF modulation that has been overlooked, with the possible exception of very short RF pulses, which are not encountered in cellular telephone technology [3-5, 9, 10]. For RF levels below the international standards (modulated or not), there are no reproducible and plausible mechanisms by which adverse biological effects can be caused in living systems. Even somewhat above guideline RF levels, the small amounts of thermal energy absorbed are within the adaptive capacity of the body and are not known to lead to disease or ill effects [5,11].

4. World Health Organization

In 1996 the World Health Organization (WHO) established the International EMF Project to assist national authorities deal with the complex EMF issues and to ensure their citizens were protected from excessive exposures from EMF systems or technologies. When I established the WHO EMF Project it was set up to:

- review the scientific literature that relates to health effects from EMF exposures;
- identify what is known and what further research is necessary to fill key gaps in knowledge so better health risk assessments can be made;
- promote high quality research;
- publish the results of research reviews using criteria that have been established by WHO over the past 60 years;
- provide information to national authorities, workers and the public about health effects and measures to protect people from EMF;
- promote international standards and protection policies that are based on sound science.

The results of the EMF Project can be found on the WHO web site at: http://www.who.int/peh-emf/en/. Since WHO started this project and recommended that specific EMF research was need to fill key gaps in knowledge, over $250 million has been spent on research throughout the world. The main purpose of this research was to determine whether biological effects could be found that resulted in detrimental consequences to health.

A set of criteria by which research results are assessed for health risk assessment reviews was published by WHO at the start of the EMF Project [12]. This paper provided guidance on the quality needed for epidemiological and animal studies, statistical analyses and importantly the standards for acceptance of scientific papers and the criteria for how health risks were to be assessed. These criteria have been used by WHO for some 60 years and are well accepted by all national and international health risk assessment committees world wide.

A series of WHO fact sheets, written in many languages, has been published on various EMF topics, mainly for public understanding of the issues and to respond to their health concerns. These can be found at: http://www.who.int/peh-emf/publications/facts/factsheets/en/index.html.

WHO staff have been criticized for not being "independent" and presumably influencing the outcomes of WHO publications. It is not possible for WHO staff to influence the conclusions and recommendations of committee reports. It is important to recognize that all conclusions and recommendations published in WHO documents are those of expert committees and WHO Task Groups convened to address particular topics. WHO staff merely facilitate the administrative procedures necessary for the task groups and committees to meet and undertake the necessary preparations for publication of committee reports. The Assistant Director General of the Cluster responsible for the work approves the membership of a WHO Task Group. They are the highest-level committees within WHO for conducting health risk assessments. Task Groups conduct a critical and thorough review of an advanced draft of the science that will be published as an Environmental Health Criteria (EHC) monograph and assess any risks to health from exposure to EMF, reach agreements by consensus, and make final conclusions and recommendations that cannot be altered after the Task Group meeting. EHC monographs and other committee reports can then be published by WHO as Fact Sheets for the general public or media, scientific publications for the scientific community and reports providing recommendations for national authorities.

Despite the huge body of scientific evidence indicating that no health effects have been established or are expected from EMF exposures at levels below the limits in the international standards, public concerns continue and have lead to calls for the adoption of precautionary measures, even though the current international standards and protective measures are already very precautionary. These public concerns have also resulted in "activist" publications, newsletters and web sites and reports by some scientists with different viewpoints; both are discussed below.

5. BioInitiative Report

An alternative view of the science was published in August 2007 [13]. The *BioInitiative Report: A Rationale for a Biologically-based Public Exposure Standard for Electromagnetic Fields (ELF and RF)*, co-edited by David O. Carpenter and Cindy Sage (authored mainly by Cindy Sage who wrote 7 of the 17 sections), has been widely circulated and has had some impact at the political level. Its stated purpose is to assess scientific evidence on health impacts from electromagnetic radiation below current public exposure limits and evaluate what changes in these limits are warranted now to reduce possible public health risks in the future.

The BioInitiative Report concludes that exposure to EMF fields contribute to a variety of diseases and subjective symptoms, including: sleeplessness, headache, fatigue, skin disorders and changes in skin sensitivity, loss of appetite, tinnitus, impairment of memory and concentration, Alzheimer's and Parkinson's disease, cardiac problems, changes in brain and nervous systems activity, stress reactions, inflammatory and allergic reactions, genotoxic effects, changes in immune system function, and many types of cancers. Yet none of these health effects has been classified as established in any national or international reviews. National and international reviews assess the whole database of scientific literature according to well-accepted international risk assessment methods and criteria (see: the European Commission's EMF-NET Coordination Action review of the BioInitiative Report [14]). Importantly the World Health Organization (WHO) [28-30], the International Commission on Non-Ionizing Radiation Protection (ICNIRP) [6] and many government and professional society reviews [10, 15-32] do not support the BioInitiative Report's conclusions.

The conclusions in the BioInitiative Report do not reflect a consensus evaluation of the research. Rather, as stated in the Preface: "information and conclusions in each chapter are the responsibilities of the authors of that chapter." Yet the Summary for the Public (section 1) states: "The clear consensus of the BioInitiative Working Group members is that the existing public safety limits are inadequate for both ELF and RF". These conflicting statements suggest that no consensus evaluation was conducted within the entire group.

The Report failed to follow appropriate and widely accepted scientific criteria for health risk assessment and criticizes the criteria used by other scientific review bodies for "adopting standards of evidence so unreasonably high as to exclude any finding of scientific concern, and thus justify retaining outdated thermal standards" without identifying any evidence to support this claim. The BioInitiative Report also argues that any effect of electromagnetic fields on biological systems should be avoided, thereby ignoring the distinctions between biological effects and adverse health effects.

Intriguingly the Report claims, "A weight-of-evidence approach was used to describe the body of evidence between health endpoints and exposure to electromagnetic fields (ELF and RF)." However individual sections give little or no evidence of assessing strengths and limitations of individual studies, quality of exposure assessment, sample size and biases (see, for example, section 6 that merely provides a catalog of studies with little or no scientific critique). One author even contradicts this statement (section 7, p. 15) by arguing that the consideration of negative and positive studies together, i.e., all of the relevant studies (essential for the weight-of-evidence method) represents an "unscientific weight of evidence approach".

The results of EMF studies cannot always be assumed to be valid. While all studies are considered, no study, positive or negative, should be accepted for health risk assessment unless it has been replicated or confirmed by independent studies. This is a WHO requirement. A key problem in assessing health risk is the lack of consistency of EMF study results. EMF research has had publications that were the result of:

- deliberate manipulation of the data;
- selection of the data so a positive outcome would result;
- basic flaws in the methodology;
- various study biases or confounders, and;
- insufficient power to determine whether an effect exists or not.

These are strong reasons not to accept any study at face value.

In the BioInitiative Report [13] the biological bases for the ELF guidelines are incorrectly characterized as relating to tissue heating. The Report claims "The safety standard assumes that EMF causes biological damage only by heating, but cell damage occurs in the absence of heating and well below the international exposure limits [6]. The stress response to both the power (ELF) and radio (RF) frequency ranges shows the inadequacy of the thermal (SAR) standard" (section 7, p. 22). The authors misunderstand the SAR dosimetry concept, and believe it to be a thermal unit. However the ELF part of the ICNIRP guidelines [6] state "...these guidelines are based on short-term, immediate health effects such as stimulation of peripheral nerves and muscles, shocks and burns caused by touching conducting objects…" (p. 496) and "Exposure to low-frequency electric and magnetic fields normally results in negligible energy absorption and no measurable temperature rise in the body" (p. 497).

Conclusions are drawn based upon a review of the data that reaches contrary conclusions. Section 2 states "Both ELF and RF exposures can be considered genotoxic (will damage DNA) under certain conditions of exposure, including exposure levels that are lower than existing safety limits" (p. 17). This apparently refers to section 7 (p. 23). Yet Drs. Xu and Chen who reviewed a similar and overlapping body of research reach a different conclusion: "To explain and/or support epidemiological observations, many laboratory studies have been conducted, but the results were controversial and no clear conclusion could be drawn to assess EMF health risk" (section 5, p. 3).

Overall the BioInitiative Report recommends precautionary measures be invoked without sound scientific evidence and then advocates mandatory measures that would be undertaken if there was a high level of agreement within the scientific community that EMF was a proven carcinogen and was effective in causing disease at very low exposure levels. It is normally accepted that precautionary measures should only be invoked when there is considerable scientific uncertainty about potential risks. In this case the BioInitiative Report dismisses all uncertainty, and hence the basis for applying any precautionary measures. While science will always contain uncertainties, no matter how much we might wish it to be otherwise, the results of some 60 years of research on EMFs have provided an excellent database from which to determine health protection measures and standards that incorporate exposure limits with a very high level of precaution in their determination.

A number of critiques of the BioInitiative Report have already been published and reach essentially the same conclusions as noted above from governmental institutions [33-36]. Perhaps the Health Council of the Netherlands [35] best summarizes these critiques when it states, "In view of the way the BioInitiative Report was compiled, the selective use of scientific data and the other shortcomings mentioned above, the Committee concludes that the BioInitiative Report is not an objective and balanced reflection of the current state of scientific knowledge. Therefore, the report does not provide any grounds for revising the current views as to the risks of exposure to electromagnetic fields."

6. Precautionary Measures

There have been calls by concerned citizens and some scientists that the Precautionary Principle should be invoked in the case of EMF exposure to the general public. A Communication from the Commission of the European Communities (CEC) [37] provided an outline of how the Precautionary Principle should be invoked. It should be considered within a structured approach to the analysis of risk that comprises three elements: risk assessment, risk management and risk communication. The CEC states "Recourse to the precautionary principle presupposes that potentially dangerous effects deriving from a phenomenon, product or process have been identified, and that scientific evaluation does not allow the risk to be determined with sufficient certainty." Clearly this is not the case with EMF since no high-level review has established that EMF exposure causes any adverse health effect below the limits in the international standards. However, if the Precautionary Principle is to be invoked, then the measures adopted should be:

- *proportional* to the chosen level of protection;
- *non-discriminatory* in their application;
- *consistent* with similar measures already taken;

- *based on an examination of the potential benefits and costs of action* or lack of action (including, where appropriate and feasible, an economic cost/benefit analysis);

- *subject to review*, in the light of new scientific data, and;

- *capable of assigning responsibility for producing the scientific* evidence necessary for a more comprehensive risk assessment.

Many scientists and concerned citizens now feel that the level of evidence for any potential harm caused by EMF exposure does not require formal application of the Precautionary Principle, but do consider that precautionary measures should be used to lower exposures below the international limits where possible. Some countries, for example, Australia, Denmark and Sweden, have adopted non-quantitative precautionary measures, but have adopted international standards as their primary basis for EMF health protection.

Adoption of precautionary measures will vary from country to country depending on their own priorities. However, many people do not realize that a high level of precaution is already built into the scientific reviews and development of standards. International exposure limits are determined assuming worst-case exposure/absorption conditions (that an individual is oriented to absorb all of the RF field that it is possible to absorb at the RF frequency), which almost never occurs. In addition, safety factors are incorporated into the limits to compensate for unknowns and uncertainties in the science. Sources of uncertainty include extrapolation of animal data to possible effects in people, differences in the susceptibility of different groups or individuals (workers versus public incl. children), statistical uncertainties in the dose-response relationship, and the uncertainty in the doses determined in individually studies.

The WHO has indicated that precautionary measures adopted by national authorities should not undermine the science-base of standards with arbitrary reduction in limits [38, 39]. If safety factors are too large, the exposure limits will be so low that some beneficial technologies may not be able to function. This however does not preclude the use of "best practice measures" as a voluntary undertaking by industry. Finally inappropriate introduction of unfounded precautionary measures can trigger concerns and amplify EMF-related risk perceptions, and may have no positive effect on trust in public health protection. In some instances they might actually prove counterproductive however well intentioned. Risk managers who intend to implement precautionary measures merely as a means for reassuring the public will probably fail [40].

7. Conclusions and the Way Forward

This paper has attempted to put the science in proper perspective and show that health risk assessment reviews must be conducted according to criteria that will reach valid conclusions and that whilst there is still much to be learnt, work must be carried out using strong scientific principles and robust research. To date, it is the considered opinion of many scientists and regulatory bodies that no well-conducted review of the whole database of scientific studies has established any health consequence from exposure to levels below the limits in the international standards. This paper concludes that the present weight of scientific evidence favours adherence to the present international standards. Since this presentation was given, ICNIRP [41] has published a statement that, following its latest review of all the scientific literature, it has reaffirmed that the exposure limits in its 1998 Guidelines [6] continue to provide a high level of public and worker protection. No research since these guidelines were published has established any health consequence below the exposure limits.

Calls for the use of precautionary measures to keep EMF exposures low while maintaining the benefits of technology are valid and reasonable given that there will always be some scientific uncertainty and that it is also good hygienic practice to keep any physical, chemical or biological agents in our living environment to levels as low as can be reasonably achieved. However, introduction of unnecessary and costly precautionary measures merely to satisfy public concerns will generally have the opposite effect. Such measures should only be introduced to address the level of scientific uncertainty that exists or if they have proven benefits.

What is the way forward? We should:

- continue EMF research to fill gaps in knowledge so better health risk assessments can be made (see WHO's RF research agenda [42]).
- conduct risk assessments, using all the scientific literature and a weight-of-evidence approach;
- adopt international standards and ensure compliance;
- help ensure that appropriate scientific principles drive and inform technological development;
- adopt realistic precautionary measures that don't undermine the science base of the standards;
- advise national authorities on facts and policies;
- disseminate results in an easily understood language so everyone can understand them;

Finally governments should establish high level scientific committees to advise on what health risks, costs and potential health benefits the scientific research indicates from proposed actions and recommendations; they must know the full facts accurately before policy development. Monitoring the recommendations of the WHO International EMF Project is essential to ensure the facts about EMF health risks are known.

References

[1] Valberg P A 2005, Modulated RF Energy: Mechanistic Viewpoint on the Health Implications. *WHO Base Station Workshop*, 15-16 June 2005, Geneva, Switzerland. http://www.who.int/peh-emf/meetings/base_stations_june05/en/index.html
[2] Valberg P A, van Deventer T E and Repacholi M H 2007, Workgroup Report: Base Stations and Wireless Networks - Radiofrequency (RF) Exposures and Health Consequences. *Environmental Health Perspectives* **115**, pp. 416–424.
[3] Challis L J 2005, Mechanisms for interaction between RF fields and biological tissue. *Bioelectromagnetics.* **26** (Supplement 7), S98-S106.
[4] Foster K R and Repacholi M H 2004, Biological effects of radiofrequency fields: does modulation matter? *Radiation Research* **162**, pp. 219-25.
[5] Sheppard A R, Swicord M L and Balzano Q 2008, Quantitative evaluations of mechanisms of radiofrequency interactions with biological molecules and processes. *Health Physics* **95**(4), pp. 365–396.
[6] ICNIRP 1998, International Commission on Non-Ionizing Radiation Protection. Guidelines for limiting exposure to time-varying electric, magnetic, and electromagnetic fields. *Health Physics* **74**, pp. 494-522. http://www.icnirp.org/documents/emfgdl.pdf
[7] Elder J A and Chou C K 2003, Auditory response to pulsed radiofrequency energy. *Bioelectromagnetics.* **24** (Supplement 6): S162-73.
[8] Aran J M, Carrere N, Chalan Y, Dulou P E, Larrieu S, Letenneur L, Veyret B and Dulon D 2004, Effects of exposure of the ear to GSM microwaves: in vivo and in vitro experimental studies. International Journal of Audiology **43**(9) pp. 545–554.
[9] Adair R K 2003, Biophysical limits on athermal effects of RF and microwave radiation. *Bioelectromagnetics* **24**, pp. 39-48.

[10] NCRP 2003, National Council for Radiation Protection "Commentary No. 18 – Biological Effects of Modulated Radiofrequency Fields." NCRP, Press Bethesda, http://www.ncrppublications.org/index.cfm?fm=Product.AddToCart&pid=4191384437
[11] Ebert S, Eom S J, Schuderer J, Apostel U, Tillmann T, Dasenbrock C and Kuster N 2005, Response, thermal regulatory threshold and thermal breakdown threshold of restrained RF-exposed mice at 905 MHz. *Physics in Medicine and Biology* **50**, pp. 5203-5215.
[12] Repacholi M H and Cardis E 1997, Criteria for EMF health risk assessment. *Radiation Protection Dosimetry* **72**, pp. 305-312.
[13] Blackman C, Blank M, Kundi M, Sage C, Carpenter D, Davanipour Z, Gee D, Hardell L, Johansson O, Lai H, Hansson Mild K, Sobel E, Xu X, Chen G and Sage S 2007, *The BioInitiative Report,* http://bioinitiative.org/report/docs/report.pdf
[14] EC FP6 Coordination Action 2007, Comments on the BioInitiative Working Group Report (BioInitiative Report), Effects of the Exposure to Electromagnetic Fields: from science to public health and safer workplace. EMF-NET Coordination Action – The Steering Committee, , http://www.izmf.de/download/archiv/EMF-NET-Bioinitiative-608.pdf
[15] Australia 2008, Australian Radiation Protection and Nuclear Safety Agency, Committee on Electromagnetic Energy. http://www.arpansa.gov.au/pubs/eme/fact1.pdf
[16] Canada 2006, Health Canada, It's Your Health, Safety and Safe Use of Mobile Phones http://www.hc-sc.gc.ca/iyh-vsv/prod/cell_e.html
[17] European Commission 2008, Health and Consumer Protection Directorate-General, Scientific Committee on Emerging and Newly Identified Health Risks (SCENIHR). Possible effects of electromagnetic fields (EMF) on human health. *Toxicology* **246**, pp. 248-250. http://ec.europa.eu/health/ph_risk/committees/04_scenihr/docs/scenihr_o_007.pdf
[18] Hong Kong 2007, Office of the Telecommunications Authority (2007) "Know More about Radiofrequency Electromagnetic Radiation" http://www.ofta.gov.hk/en/freq-spec/radiation.pdf
[19] IEGMP 2000, Independent Expert Group on Mobile Phones Mobile phones and health. Report of an Independent Expert Group on Mobile Phones. Chairman, Sir William Stewart. Chilton, NRPB http://www.iegmp.org.uk/
[20] Institution of Engineering and Technology 2006, Biological Effects Policy Advisory Group on Low-level Electromagnetic Fields, The Possible Harmful Biological Effects of Low-Level Electromagnetic Fields of Frequencies up to 300 GHz. http://www.theiet.org/publicaffairs/bepag/postat02final.pdf
[21] Ireland 2007, Expert Group on Health Effects of Electromagnetic Fields. http://www.dcmnr.gov.ie/NR/rdonlyres/9E29937F-1A27-4A16-A8C3-F403A623300C/0/ElectromagneticReport.pdf
[22] Japan 2007, Ministry of Internal Affairs and Communications, Committee to Promote Research on the Possible Biological Effects of Electromagnetic Fields http://www.soumu.go.jp/joho_tsusin/eng/Releases/NewsLetter/Vol18/Vol18_06/Vol18_06.pdf
[23] Netherlands 2007, Health Council of the Netherlands http://www.healthcouncil.nl/pdf/Press%20release%20200706%20site.pdf
[24] New Zealand 2007, Ministry of Health, National Radiation Laboratory. Safety of Cell Phones. http://www.nrl.moh.govt.nz/faq/cellphonesandcellsites.asp
[25] Royal Society of Canada 2004, Recent Advances in Research on Radiofrequency Fields and Health: 2001-2003 http://www.rsc.ca/index.php?page=expert_panels_rf&lang_id=1&page_id=120
[26] Sweden 2008, Swedish radiation protection agency, Recent Research on EMF and Health Risks. Fifth Annual Report from SSI:s Independent Expert Group on Electromagnetic fields, 2007 Revised edition 15 April, 2008. Stockholm, Sweden http://www.stralsakerhetsmyndigheten.se/Global/Publikationer/Rapport/Strålskydd/2008/ssi-rapp-2008-12.pdf
[27] USA 2006, Federal Communications Commission Mobile Phones and Health Concerns http://ftp.fcc.gov/cgb/consumerfacts/mobilephone.html
[28] USA 2008, US Food and Drug Administration/CDRH: Cell Phones - Reducing Exposure: Hands-free Kits and Other Accessories. Available on web site at: http://www.fda.gov/cdrh/wireless/health-exposure.html (Update 14 October 2008).

IOP Conf. Series: Earth and Environmental Science **10** (2010) 012001 doi:10.1088/1755-1315/10/1/012001

[29] USA 2008a, National Cancer Institute. Cell Phones and Brain Cancer: What We Know (and Don't Know). *NCI Cancer Bulletin* **5**(19), pp. 7-8.

[30] USA 2008b, National Cancer Institute Fact Sheet Cellular Telephone Use and Cancer Risk. http://www.cancer.gov/cancertopics/factsheet/Risk/cellphones

[31] UK 2004, National Radiation Protection Board. Review of the Scientific Evidence for Limiting Exposure to Electromagnetic Fields (0 – 300 GHz). Documents of the NRPB, **15**(3). http://www.hpa.org.uk/radiation/publications/documents_of_nrpb/pdfs/doc_15_3.pdf

[32] UK 2004a, National Radiation Protection Board. Mobile phones and health 2004. Report by the Board of NRPB. Documents of the NRPB, 15(5) http://www.hpa.org.uk/radiation/publications/documents_of_nrpb/abstracts/absd15-5.htm

[33] Denmark 2007, National Board of Health. Comments on BioInitiative Report. http://www.sst.dk/upload/forebyggelse/cff/miljoemedicin/ikke_ion/bioinitiative_4okt07.pdf - *in Danish.*

[34] Germany 2008, Bundesamt für Strahlenschutz (Federal Office for Radiation Protection, BfS). Comments on the Bioinitiative report. http://www.bfs.de/de/elektro/papiere/Synopse_EMF.pdf - *in German.*

[35] Netherlands 2008, http://www.gr.nl/pdf.php?ID=1743&p=1

[36] European Commission 2007, Comments on the BioInitiative Report EMF-NET Coordination Action http://emf-net.isib.cnr.it

[37] Commission of the European Communities 2000, Communication from the Commission on the precautionary principle. Brussels. See: http://ec.europa.eu/dgs/health_consumer/library/pub/pub07_en.pdf

[38] WHO 2000, Fact Sheet #193 Mobile phones and base stations. http://www.who.int/mediacentre/factsheets/fs193/en/

[39] WHO 2007, Fact Sheet #322, Exposure to extremely low frequency fields. http://www.who.int/mediacentre/factsheets/fs322/en/index.html

[40] Weidemann P M, Thalmann A T, Grutsch and Schütz H 2006, The impacts of precautionary measures and the disclosure of scientific uncertainty on EMF risk perception and trust. *Journal of Risk Research* **9**(4), pp. 361-372.

[41] International Commission on Non-Ionizing Radiation Protection 2009, ICNIRP statement on the "Guidelines for limiting exposure to time-varying electric, magnetic, and electromagnetic fields (up to 300 GHz)" *Health Physics* **97**(3), pp. 257-258. http://www.icnirp.de/documents/StatementEMF.pdf

[42] WHO 2006, 2006 WHO Research Agenda for Radio Frequency Fields, World Health Organisation http://www.who.int/peh-emf/research/rf_research_agenda_2006.pdf)

IOP Conf. Series: Earth and Environmental Science **10** (2010) 012002 doi:10.1088/1755-1315/10/1/012002

Methodological Approach to EMF Protection Standards

Paolo Vecchia

Department of Technology and Health, National Institute of Health, Rome, Italy

Email: paolo.vecchia@iss.infn.it

Abstract. Biological and health effects of electromagnetic fields (EMF) have been investigated for many years. Exposure standards have been developed internationally, that provide adequate protection against all known adverse effects of exposure to EMF. The guidelines developed by the International Commission on Non Ionizing Radiation Protection (ICNIRP) are widely recognized and have formed the basis for national regulations in several countries. Concerns for hypothesized, but unproven, long-term effects of chronic exposure to low-level EMFs have created a demand for precautionary measures beyond the standards for recognized, acute effects. However, such measure should be separate from exposure standards, and adopted with special care to avoid undermining the credibility of science-based guidelines, and of health protection agencies.

1. Introduction

Recommendations for the safe exposure of workers and the general public to electromagnetic fields (EMF) have been issued by a number of international organizations, as well as scientific institutions and national health authorities in several countries. Although some difference exists in the numerical values of recommended limits and other details, all the standards are based on the same scientific database, and on the same basic approach.

These standards have been widely accepted, and an increasing number of countries have adopted the international guidelines as their own national regulations. However in recent years criticism has been expressed by activist groups, media, and also some scientists claiming that existing standards do not adequately protect against a variety of potentially adverse effects that some studies have suggested as a long-term response to chronic exposures to low level electromagnetic fields.

The controversies focus on two main arguments: on the one side, there are diverging opinions on the degree of evidence of such effects, and therefore on the credibility that they actually exist; on the other side, it is disputed whether, and how, protection standards should address effects that have been hypothesized, but not adequately proved by scientific research.

In this context, it seems important to clarify how science-based protection standards are developed. To this purpose, reference is made in particular to the exposure guidelines issued by the International Commission on Non Ionizing Radiation Protection (ICNIRP). As already mentioned, however, most of the considerations developed in these notes equally apply to standards developed by other scientific bodies and based on the same philosophy.

2. Structure and Role of ICNIRP

ICNIRP is an independent group of expert to evaluate the scientific literature on biological effects of non ionizing radiation (that includes electromagnetic fields, optical radiation, ultrasound and infrasound), to assess possible health risks, and to provide advice to national governments and international organizations on possible ways to protect against such risks. It was created in 1992, to replace an International Non Ionizing Radiation Committee (INIRC) that operated from 1977 to 1992 inside the International Radiation Protection Association (IRPA). ICNIRP took over tasks and responsibilities of IRPA/INIRC, with no solution of continuity; therefore, the Commission represents at present over thirty years of experience in the field of Non Ionizing Radiation (NIR) protection.

ICNIRP is composed of 12 members, plus a Chairperson and a Vice-Chairperson. The complexity of the issues requires a multi-disciplinary approach, and ICNIRP membership includes experts in different areas such as medicine, biology, toxicology, epidemiology, physics and engineering. To provide a wider representation of these disciplines, and to assist the Main Commission in its tasks, Standing Committees exist on Epidemiology, Biology, and Physics and Engineering, respectively. A fourth Standing Committee is specialised on Optical radiation. Over the years, ICNIRP has gained a solid reputation: it is formally recognized by international bodies such as the World Health Organization (WHO) and the International Labour Office (ILO). It also collaborates with a number of other institutions, including the Commission of the European Union (EC).

Among the activities of ICNIRP, it is of special importance the development of guidelines for the safe exposure of workers and the general public are periodically revised in the light of the advancement of research. Guidelines for static magnetic fields were published in 1994 [1], while the last revision of guidelines for time-varying electric, magnetic, and electromagnetic fields of frequencies up to 300 GHz was issued in 1998 [2]. New revisions are in progress for static magnetic fields, and for low frequency (up to 100 kHz) electric and magnetic fields The publication of both is expected in 2009, while a revision of the guidelines on radiofrequency fields will be carried out in the more distant future, after the completion of some important epidemiological and biological studies that are presently ongoing, and a subsequent, comprehensive risk assessment.

The ICNIRP guidelines have gained a large credit, and have constituted the basis for national regulations in over thirty countries around the word. The European Union issued in 1999 a Recommendation for the protection of the general public [3], and in 2004 a Directive for the safe exposure of workers [4], that completely endorse the ICNIRP recommendations.

A basic feature of the ICNIRP standards, as well as any other of ICNIRP's guidance and advice, is that they are firmly based on established science, and aim at protecting against all – and only – adverse effects that have been clearly indicated by qualified research.

In recent years, however, a culture of precaution has progressively emerged, in all fields of environmental and health protection. Consequently, the demand has increased for policies that go beyond the prevention of established effects, taking in some consideration also partial or preliminary results of the research, and health risks that have not been definitely established. This has led to a broader perspective of health protection, with consideration to other factors than scientific findings, such as socioeconomic implications of health policies. Different systems of protections have been developed, that may be alternative or complementary to one another.

3. The Systems of Protection

Different systems of protection are generally adopted for different situations, depending on the nature of the effects and the quality of scientific data. A schematic distinction can be made between:

- *Health threshold based systems* that are adequate when biological effects that might lead to health detriment have been established, and thresholds for them have been identified. The protection of physical health is provided through exposure limits (or dose limits, depending on the nature of the agent), to assure that exposures are below the thresholds. Such approach allows, in principle, the total prevention of the identified adverse effects.

- *Optimization systems* that may be appropriate in face of a known and accepted hazard, where a threshold cannot be established. This is typical of established effects that are stochastic in nature. The knowledge of the hazard includes the identification of a dose-response relationship, that is most frequently monotonic, with health risk reducing to zero at zero exposure. Rather than preventing adverse effects, such systems aim at defining – in an objective way – the most acceptable level of risk, i.e. the best balance of costs and benefits of measures adopted to reduce the health detriment. A well-known example is the ALARA (As Low As Reasonably Achievable) principle adopted in the area of ionizing radiation.

- *Precautionary measures* that may be adopted in case of uncertainty, i.e. to protect against hazards that have been suggested, but not established by scientific research. Most frequently, these measures are implemented - or invoked - in observance of the Precautionary Principle.

While the two latter systems require economical, social and political considerations to be taken into account, all the three must be based on solid and reliable scientific data. The starting point for the selection, the development, and the implementation of any protection system is therefore an in-depth analysis of the literature, and a scientific assessment of health risks. ICNIRP performs this task in cooperation with other international bodies, in particular the World Health Organization (WHO) and the International Agency for Research on Cancer (IARC).

4. Scientific Assessment of Risk

Tight links have been created between international bodies in order to avoid redundant activities and create the most effective synergies. In particular, a specific sequence of actions has been established in order to provide to authorities, workers, and the public the best possible advice on all health issues related to EMF.

On commitment by WHO, ICNIRP carries out a comprehensive review of the scientific literature concerning exposure assessment and dosimetry, biological effects, and epidemiology. On its side, IARC evaluates the available data specific to a possible role of EMF in the development of cancer, with the final aim of classifying the different types of EMF according to their carcinogenic power. With these inputs, WHO performs a global evaluation of all possible health risks of EMF exposure. Finally, ICNIRP uses the conclusions of all the above steps to revise and update its guidelines as appropriate.

The actions are therefore complementary. However, each organization has its own criteria, depending also on its mission. The basic approach adopted by ICNIRP is detailed in an *ad hoc* scientific paper [5].

The review of the literature is at the same time comprehensive and selective. On the one side, all the research body is examined, and not only the most recent findings. On the other, a rigorous selection is made based on the scientific quality of each paper. As for any other field of science, publication in a recognized journal following a peer review is a necessary condition, although not necessarily sufficient, for the paper to be included in the database for the assessment of biological and health effects.

In such assessment three steps can schematically be identified:

- Initially, each study is evaluated in terms of its relevance for the health effect being considered, and the quality of methods used. Different weights may be assigned to the studies, depending on the extent to which they meet quality criteria regarding e.g. the experimental techniques used, the assessment of exposure, the control of experimental conditions, possible biases and confounders, the replicability of the experiments and the reproducibility of the results.

- As a second step, all information relevant for each health effect is evaluated. This review is normally carried out separately for epidemiological investigations, human laboratory tests, animal studies, and *in vitro* research.

- Finally, the outcomes of the above steps are combined in an overall evaluation, taking the consistency of data into proper consideration. ICNIRP recognizes that this process involves some judgements; however, the collective participation to the process minimizes bias due to personal attitudes.

When the overall evaluation allows the identification of an effect that is causally related to the exposure, the effect becomes *established.* Leading criteria in the identification of effects are the reproducibility of findings, and the consistency across studies of different nature (e.g. data from laboratory research *in vitro* and *in vivo* that may give biological plausibility to a causal interpretation of statistical correlations when these are indicated by epidemiological studies).

The established effects shall be quantitatively related to the exposure. However, the entity of a given effect does not depend only on the level of the external field, but also on the coupling of the field with the exposed body, or selected organs. The quantitative relationship by which the external exposure affects a biologically effective parameter of the target tissue is unique to a single exposure condition. Therefore, effects are better described by quantities that reflect the efficacy by which the external exposure causes a certain biological effect. These are termed *biologically effective quantities*, or *dosimetric quantities*.

Different biologically effective quantities have been identified as appropriate for different interaction mechanisms and biological effects. In general – but not always – these quantities are internal to the body and therefore cannot be directly measured. A correspondence shall therefore be established between biologically effective quantities and external fields, taking exposure conditions in due account. This is accomplished through theoretical and experimental modelling techniques that constitute what is called *dosimetry*, in analogy with toxicology and ionizing radiation.

By means of biologically effective quantities, established adverse effects can generally be ranked according to the exposure level at which each effect becomes relevant. The effect that is relevant at the lowest level of exposure is called the *critical effect*, and is the criterion for the definition of exposure limits. The limitation of exposure to levels below the thresholds for the critical effect provides, *a fortiori*, protection against any other adverse effect.

It should be noted that in this process the different sensitivities, and the ability to tolerate EMFs, of different groups of the population are taken into account. The critical effect is selected with special consideration to categories that might exhibit lower tolerance, including children, the elderly, and some chronically ill people. The guidelines are therefore adequate to protect all the population groups, to the extent to which the corresponding scientific knowledge is adequate.

5. Basic Restrictions and Reference Levels

A distinctive feature of the ICNIRP guidelines – common to other international standards - is the two level structure used. The biological and health effects depend on several parameters that characterize exposure. *Basic restrictions* are defined in terms of the appropriate biologically effective quantities, and are set below the threshold for the appropriate critical effects. Due to practical difficulties in measuring or calculating some biologically effective quantities, these are related to *reference levels* that are expressed in terms of a directly measurable parameter of the external exposure.

The strategy is conservative. The use of reference levels assures in fact compliance with the basic restrictions, since the relationships between them have been developed for conditions of maximum coupling between the external fields and the exposed person. On the other hand, exceeding the reference levels does not necessarily imply that the basic restriction is exceeded; whether this occurs or not should be ascertained through a more detailed investigation.

Both basic restrictions and reference values are affected by uncertainties, due to the intrinsic variability of biological data, experimental errors, uncertainties in the extrapolation of animal data to humans, limitation in dosimetry, biases and confounders. Reduction factors are therefore conservatively introduced, whose magnitude varies depending on the degree of incertitude.

To avoid possible misunderstandings, it shall be clarified that reduction factors are not intended to compensate for gaps in knowledge. In effect, their use as a precautionary measure to account for uncertainty in science has been criticized as inappropriate by standard-setting bodies and health protection agencies. WHO notes that *"science-based exposure limits should not be undermined by the adoption of arbitrary cautionary approaches. That would occur, for example, if limit values were lowered to levels that bear no relationship to the established hazards or have inappropriate arbitrary adjustments to the limit values to account for the extent of scientific uncertainty"* [6].

6. Health Effects of Electromagnetic Fields

The extensive research carried out in particular in the last decades indicates that established effects are acute, in the sense that they occur immediately, or within a short time, after the exposure. Such acute effects exhibit thresholds, i.e. they become apparent only above given exposure levels.

Not only are such acute effects supported by strong laboratory evidence, but the basic interaction mechanisms behind them are clearly understood. At extremely low frequencies (ELF) the mechanism is stimulation of electrically excitable tissues such as muscles and nerves, while in the range of radiofrequency (RF) and microwaves (MW) the electromagnetic energy impinging on the exposed body is absorbed by the biological tissues and dissipated as heat, causing a general or local increase of the body temperature. The resulting biological and health effects are related to the temperature increase rather than electromagnetic fields per se. Because of this, they are generally termed thermal effects.

Comprehensive reviews of the biological and health effects of different kinds of electromagnetic fields can be found in several reports, and in particular in monographs published by WHO in the series Environmental Health Criteria (EHC) Documents [7-9].

7. Precautionary Policies

The international guidelines are aimed at preventing all the health effects that have been scientifically established using the criteria discussed in previous sections. Such effects are acute, and are related to the stimulation of electrically excitable tissues or to the absorption of electromagnetic, for fields of low- or high-frequency, respectively.

However, the possibility of long-term effects of chronic exposure below the thresholds for acute effects cannot be dismissed, and extremely low frequency (ELF) magnetic fields have been classified by IARC as "possibly carcinogenic to humans" (group 2B) [10]. To face these possible risks, some national governments, or local authorities, have adopted precautionary policies, to replace or complement science-based exposure limits. Quite often, the *Precautionary Principle* is invoked to this purpose.

In spite of its popularity, the principle is not well defined, and is variously interpreted. In addition, a possible conflict between science and the principle has been outlined [11]. An important clarification was provided by the European Commission (EC) [12]; it stressed that a basic condition for the principle to be invoked is that a potentially serious health hazard had been identified and scientifically evaluated. Therefore, science is a fundamental basis – though not the only one – for the adoption of precautionary measures.

Other criteria are indicated by the EC for the correct application of the principle. The selected measures should be *inter alia*:

- tailored to the chosen level of protection;
- non-discriminatory, i.e. treat comparable situations in a similar way;
- comparable to measures already taken in equivalent areas;
- based on an examination of the potential benefits and costs;
- provisional, i.e. subject to review in the light of new scientific data.

Examining in this respect the case of EMF, WHO considers that *"[...] a cautionary policy for EMF should be adopted only with great care and deliberation. The requirements for such a policy as outlined by the European Commission do not appear to be met in the case of either power or radio frequency EMF"* [5]. Indeed, no health risk has been clearly identified for radiofrequency fields, where a number of recent studies have focused on possible long term effects of mobile phone use. The concern for possible brain tumours appears to be related to the fact that the antenna is very close to the head, rather than to scientific findings, that for the large majority have been negative.

In the case of ELF fields, where a risk extremely severe in its nature (childhood leukaemia) has been suggested by epidemiological studies, the criterion of risk/benefit is generally not met. An analysis carried out by an *ad hoc* commission in Italy [8] indicated that the nation-wide reduction of exposure to magnetic fields generated by power lines to no more that 0.5 μT would imply a total cost of over 20,000 million euros, corresponding to about 1,000 million euros per "statistical life" saved in the unproved hypothesis of carcinogenicity of magnetic fields. This amount largely exceeds the "willingness to pay" of any society.

The inapplicability of the precautionary principle does not necessarily mean disregarding any precaution. On the contrary, it has been suggested that in the presence of scientific incertitude (unavoidable) any political decision be taken in the context of a *precautionary framework*, where social and economic factors, including the sensitivities of the public, are taken into account besides scientific evidence of risk.

In this context, it is also important to put the health risks of EMF in the appropriate perspective. It is worth to note, for example, that the issue has received very limited attention in comprehensive reviews of IARC [13] and WHO [14] on cancer and on children's health, respectively.

In 2007, WHO issued a comprehensive report on possible risks of ELF electric and magnetic fields [9]. The conclusions are illustrative of the above considerations:

"In recommending precautionary approaches, an overriding principle is that any actions taken should not compromise the essential health, social and economic benefits of electric power. In the light of the current scientific evidence and given the important remaining uncertainties, it is recommended that an assessment be conducted of the impact of any precautionary approach on the health, social and economic benefits of electric power. Provided that these benefits are not compromised, implementing precautionary procedures to reduce exposures is reasonable and warranted. The costs of implementing exposure reductions will vary from one country to another, making it very difficult to provide a general recommendation for balancing the costs against the risk from ELF fields. Given the weakness of the evidence for a link between exposure to ELF magnetic fields and childhood leukaemia and the limited potential impact on public health, the benefits of exposure reduction on health are unclear and thus the cost of reducing exposure should be very low".

8. Conclusions

Comprehensive systems of protection have been developed at the international level, and adopted in a large number of countries. They are conservative, flexible, and based on solid science, so providing adequate protection against all known health effects of EMF.

In response to the concerns of the public, and given some uncertainties that still exist in some areas of scientific knowledge, consideration of precautionary measures could be warranted in some cases. A basic requirement, however, is that these measures are adopted in such a way as not to undermine the credibility of the international standards, and consequently the trust in health authorities and in science.

References

[1] ICNIRP 1994, Guidelines on Limits of Exposure to Static Magnetic Fields. *Health Physics* **66**, pp. 100-106.

[2] ICNIRP 1998, Guidelines for limiting exposure to time varying electric, magnetic, and electromagnetic fields (up to 300 GHz). *Health Physics* **74**, pp. 494-522.

[3] EU 1999, Council Recommendation on the limitation of exposure of the general public to electromagnetic fields 0 Hz - 300 GHz. (Official Journal of European Communities L 199/59 of 30 July 1999).

[4] EU 2004, Corrigendum to Directive 2004/40/EC of the European Parliament and of the Council of 29 April 2004 on the minimum health and safety requirements regarding the exposure of workers to the risks arising from physical agents (electromagnetic fields) (Official Journal of the European Union L 159 of 30 April 2004).

[5] ICNIRP 2002, General approach to protection against non-ionizing radiation. *Health Phys.* **82**, pp. 540-548.

[6] WHO 2000, *Electromagnetic fields and health: Cautionary policies.* www.who.int/docstore/peh-emf/publications/facts_press/EMF-Precaution.htm

[7] UNEP/WHO/IRPA 1981, Environmental Health Criteria 137. Electromagnetic Fields (300 Hz – 300 GHz). Geneva: WHO. http://www.inchem.org/documents/ehc/ehc/ehc137.htm

[8] WHO/ICNIRP/ILO 2006, Environmental Health Criteria 232. Static Fields. Geneva: WHO. http://www.who.int/entity/peh-emf/publications/reports/ehcstatic/en/index.html

[9] WHO/ICNIRP/ILO 2007, Environmental Health Criteria 238. Extremely Low frequency (ELF) Fields. Geneva: WHO. http://www.who.int/pehemf/publications/elf_ehc/en/index.html

[10] IARC 2002, Monographs on the Evaluation of Carcinogenic Risk to Humans, **80**. Nonionising Radiation Part 1: Static and Extremely Low Frequency (ELF) Electric and Magnetic Fields. Lyon: International Agency for Research on Cancer.

[11] Foster K R, Vecchia P and Repacholi M H 2000, Science and the precautionary principle. *Science* **288**, pp. 979-981.

[12] European Commission 2000, Communication on the Precautionary Principle. http://europa.eu.int/comm/off/com/health_consumer/precaution.htm

[13] IARC 2003, World Cancer Report (Stewart B W and Kleihues P, Eds.). Lyon: International Agency for Research on Cancer.

[14] WHO/Regional Office for Europe and European Environmental Agency 2002, Children's health and environment: A review of evidence (Tamburlini G, v. Ehrenstein O and Bertollini R, Eds.). Luxembourg: Office for Official Publications of the European Communities.

IOP Conf. Series: Earth and Environmental Science **10** (2010) 012003 doi:10.1088/1755-1315/10/1/012003

Electromagnetic Fields and the Public: EMF Standards and Estimation of Risk

Yury Grigoriev

Russian National Committee on Non-Ionizing Radiation Protection, RNCNIRP, and Federal Medical Biophysical Centre, FMBA, Moscow, Russia.

Email: profgrig@rambler.ru

Abstract. Mobile communications are a relatively new and additional source of electromagnetic exposure for the population. Standard daily mobile-phone use is known to increase RF-EMF (radiofrequency electromagnetic field) exposure to the brains of users of all ages, whilst mobile-phone base stations, and base station units for cordless phones, can regularly increase the exposures of large numbers of the population to RF-EMF radiation in everyday life. The need to determine appropriate standards stipulating the maximum acceptable short-term and long-term RF-EMF levels encountered by the public, and set such levels as general guidelines, is of great importance in order to help preserve the general public's health and that of the next generation of humanity.

1. Introduction

Electromagnetic conditions in the environment have changed greatly within the past 15 years. This is in part due to the introduction of mobile telecommunication systems, which have essentially changed the degree of electromagnetic exposure members of the general public are exposed regularly to.

Radiation from mobile-phone communications devices - *and others forms of cordless communications technology* - present completely new electromagnetic exposure regimes for the general population. As examples of this, the daily use of mobile phones is shown to increase RF-EMF (radiofrequency electromagnetic field) exposure to users' brains unless special precautions are taken, and the operation of wireless devices and mobile-phone base-stations can dramatically increase the exposures of large portions of the public to RF-EMF radiation.

The requirement to develop appropriate standards stipulating maximum acceptable short-term and long-term RF-EMF exposure levels for the general public (and it is argued, for vulnerable subsets of the public such as children and pregnant women), and set such levels as general guidelines appears of great importance particularly when technological advances are being made so quickly. Such measures promptly undertaken can help ensure the development of appropriate communications technology whilst safeguarding the general public's health and that of future generations. We do not have the right to make mistakes on such a major issue through complacency.

2. Characteristic Conceptions of Russian and Foreign Standards

For the development of appropriate standards that take health and wellbeing into account; it appears necessary to formulate the hygiene hypothesis as follows:

"*The hygienic standards are for the protection of the population, taking into account factors potentially harmful to health, and with the obligation of taking into account typical prevalence of these factors in the general population*" [1].

2.1 International Standards

It is important to consider existing international standards and how these conform to this postulate. Presently international standards are determined by ICNIRP, IEEE, CENELEC and other national and international commissions. The first three organisations mentioned determined the basis for their standards some years ago, and continue to propose RF-EMF guidelines that only take into account thermal, acute and pathological effects [2, 3, 4]. Unfortunately, these RF-EMF standards do little to provide protection for the general population.

With regard to international standards and likely exposures, it is imperative to recognise that the general population is not typically exposed to thermal levels in the workplace or in everyday life. The establishment of a proposed threshold level for pathological effects presently makes the assumption that compensative or adaptive reactions will occur in an organism. However we strongly disagree with this assumption. Whilst people very rarely receive acute exposures in everyday life, all populations in the world have are chronically exposed on a daily basis to low levels of RF-EMF that can cause biological effects and standards have to be set accordingly to take this into account [5].

2.2 National Standards

The RF-EMF exposure levels deemed acceptable vary greatly between countries. As noted in the BioInitiative Report [6] for the RF frequency range of 800-900 MHz, exposure levels range from 4 $\mu W/cm^2$ (0.04 W/m^2) in Switzerland, 6 $\mu W/cm^2$ (0.06 W/m^2) in China, and 10 $\mu W/cm^2$ (0.1 W/m^2) in Russia and Italy to 580 $\mu W/cm^2$ (5.8 W/m^2) in the United States and 5800 $\mu W/cm^2$ (58.0 W/m^2) in the United Kingdom.

2.3 The Russian RF-EMF standards

First of all, it is necessary to provide background information to the reader on the Russian National Committee on Non-Ionizing Radiation Protection (RNCNIRP). The RNCNIRP was created in 1997 at the Russian Academy of Medical Science (RAMS) within the framework of the Russian Scientific Commission on Radiation Protection (RSCRP). The RSCRP acts as the overseer of the RNCNIRP.

The RNCNIRP employs 40 specialists, of whom 38 are qualified scientists, and 2 members are representatives of the Ministry of Health. The RNCNIRP is an independent scientific organisation which does not accept financial sponsorship. Its decisions are thought of as recommendations, and are considered by the Ministry of Health of the Russian Federation when it is setting standards [7].

Mandatory compliance is required with regard to the Sanitary Provisions and Ecological Norms (SanPiN) guidelines set by the Ministry of Health of the Russian Federation. The latest RF-EMF SanPiN 2.1.8/2.2.4.1190-03 (safety standard) on mobile communications was issued by the Ministry of Health of the Russian Federation in 2003. This decreed that the maximum permissible exposure level for RF-EMF over the frequency range of 300 MHz – 300 GHz in the Russian Federation is 10 $\mu W/cm^2$ (0.1 W/m^2). This SanPiN also recommended that:

"*Use of mobile telecommunication devices should be restricted for those under 18 years of age and pregnant women*" [8].

3. Limitations of Present Standards in Russia and Abroad

There are no publications that present ways of extrapolating from the various existing standards recommendations to properly assess real environmental conditions for the population. There are also currently no proposals on how to estimate danger by using existing international standards recommendations: from acute influences to chronic exposure, and from thermal levels to non-thermal levels.

The methodology used to create standards in Russia (and formerly the USSR, the former Soviet Union) uses principles based on additional factors found during actual EMF exposure of the population. These take into account:

- non-thermal levels;
- chronic exposures;
- establishment of 'working levels', instead of 'threshold levels'.

Consideration is also taken into account of the presence of adaptation processes from chronic exposures instead of direct pathological effects.

4. Database for Standardisation

The results of research undertaken in industrial settings from 1950 onwards – almost 60 years of data – and chronic exposure experiments, etc., provide the database used for Russian (and formerly USSR) standardisation. The necessity of developing standards for children as new risk group is also now recognised, as is the need for need for collecting data on how they are affected by typical exposures.

The potential health risk for children is very high and creates a completely new problem we urgently need to address. As noted by the WHO:

"*Children are different from adults. Children have a unique vulnerability. As they grow and develop, there are 'windows of susceptibility': periods when their organs and systems may be particularly sensitive to the effect of certain environmental threats*" [9].

Modern children will generally use mobile phones for a longer overall period than adults of the present generation because they have started to use them at earlier age and will continue use such technology when they become adults. For a variety of reasons it is impossible to use data obtained on adults as an accurate predictor for effects on children. It is therefore necessary to develop standards which take into consideration localised brain exposures as related to age and undertake corresponding research. There are presently no studies investigating the effects of chronic RF-EMF exposure to the brain area which take into account investigation of the possible effects of such exposures in the developing brain an area we hope to address in future work.

5. Thermal and Non-Thermal Health Effects

Presently there is a refusal by many western scientists to accept the possibility of the existence of detrimental non-thermal RF-EMF effects, which in consequence has resulted in an under estimation of the actual dangers that may exist to the health of the population through different degrees of RF-exposure. However there are a very large number of publications by scientists from both the east and the west documenting biological effects from low levels of RF-EMF.

As examples of this, reference is made to the following: the BioInitiative Report which was authored by 14 respected scientists from five countries (Austria, China, Denmark, Sweden and the USA) [6], the Stewart Report in the United Kingdom [10], and the US National Academy of Sciences National Research Council [11].

Numerous peer-reviewed studies covering non-thermal biological effects have been made by scientists in Russia (previously the USSR) [12-14]. Analysis of 28 biological experiments conducted *in vitro*, *in situ*, and *in vivo* by the present author from 1975 onwards in the former Soviet Union, and later in Russia, using modulated RF-EMF allows the following basic conclusions to be made:

- exposure of biological systems to EMF with higher or lower composite regimens of modulation can lead to the possible development of both physiological and unfavourable biological effects, that are distinct from the biological effects induced by non-modulated EMF;

- acute exposure to low intensities of modulated EMF (at non-thermal levels) can result in development of pathological effects;

- there is a dependence of development of a reciprocal biological response on the intensity and directness of the concrete regimen of EMF modulation; this dependence was fixed at all levels of biological systems — *in vitro*, *in situ*, and *in vivo*;

- as a rule, modulated EMF invokes more recognisable biological effects than continuous EMF regimes.

6. Unsolved Problems in Estimating Future RF-EMF Risk

With the creation of new conditions new potential problems arise:

- Problems of accumulation of effect, remote somatic effects and cancer.
- Problems of adaptation.
- Estimation of the influence of simultaneous exposure to various frequencies.
- Estimation of the role of signal modulation.
- Coordination of the criterion to establish a threshold or appropriate 'working level'.
- Changing reactivity and appearance of electromagnetic hypersensitivity.
- Modulation and biological effects.

Such challenges indicate the benefits of extending research into the biological effects of RF-EMF as early as possible when potential new conditions arise.

7. Guaranteeing Good Health for the Population – A Discussion

The present scientific thinking and basis used in many instances for developing appropriate RF-EMF safety standards does not correspond realistically to modern conditions of RF-EMF exposure as experienced by members of the public (both for generalized exposures and through direct use of mobile communication systems technologies).

- From what we now know, existing safety standards (both foreign and Russian) have become outdated. Modern accumulative RF-EMF exposures have also increased considerably from those found in the past, thereby further increasing likely risk.

- The existing standards cannot guarantee the safe, healthy development of the next generation of humanity.

The viewpoint of the Mobile Manufacturers Forum (MMF) - an international association of radio communications equipment manufacturers – differs from that of ourselves and is as follows:

"*the MMF believes that there is a strong scientific basis for all consumers to have confidence in the safety of mobile phones and base stations. In addition, we fully support parents deciding for themselves whether they want their children to use a mobile phone or not.*" [15].

We make no comment on this large divergence in viewpoint from that of our own researchers and ask you to make your own assumptions as to why there may be such a difference in opinions!

8. Our Conclusions

We propose that it is of the greatest necessity to undertake appropriate detailed research and accumulate suitable knowledge for preparing proper precautionary standards based on the best available scientific evidence at the earliest possible opportunity.

As part of this work it would be advisable to:

- study the possible effects of repeated RF-EMF exposures from mobile phone use (and exposure to RF-emitters) over periods of several years on the brains of children, teenagers and adult users. It is proposed that the childhood studies should monitor subjects from the age of seven onwards;

- develop and undertake new long-term standardization measures, including measures related directly to suitable exposure levels for children;

- put forward more rigid requirements for industries using technologies operating over such frequency ranges;

- actively introduce, and promote, the adoption of ***proper*** precautionary principles.

The thesis held by some that the present forms of RF communications are absolutely safe is both premature and potentially dangerous. It also holds back the development of more appropriate technologies.

It is necessary to educate scientists, politicians, industries and the general public, including parents and children, that mobile communication devices (and hands-free devices) are not toys, and should be used carefully in a responsible manner. There is no room, or time, for complacency on these matters as related to health, and adopting such measures sooner rather than later could prove highly advantageous to all parties concerned.

References

[1] Grigoriev Y 2008, Russian NCNIRP Guidance. New condition of EMF RF exposure and guarantees of population health. Presentation given at *EMF & Health - A Global Issue: Exploring appropriate precautionary approaches,* The Radiation Research Trust Conference, The Royal Society, London, United Kingdom. 8 - 9 September, 2008. Abstract available at: http://www.radiationresearch.org/conference/downloads/021235_grigoriev.pdf

[2] ICNIRP 1998 ICNIRP (International Commission on Non-Ionising Radiation Protection) ICNIRP Guidelines, Guidelines for limiting EMF Exposure to time-varying electric, magnetic, and electromagnetic fields (up to 300 GHz), *Health Physics*, **74** (4), pp. 494-522.

[3] IEEE Standard C95.1-2005, IEEE Standard for Safety Levels with Respect to Human Exposure to Radio Frequency Electromagnetic Fields, 3 kHz to 300 GHz http://standards.ieee.org/reading/ieee/interp/C95.1-2005.html

[4] CENELEC 2002, Basic Standard for the Measurement of Specific Absorption Rate Related to Human Exposure to Electromagnetic Fields from Mobile Phones (300 MHz-3 GHz), CENELEC EN 50361, European Committee for Electrotechnical Standardization / Comité Européen de Normalisation Electrotechnique / Europäichsches Komitee für elektrotechnische Normung (CENELEC).

[5] Grigoriev Y, Vasin A and Grigoriev O 2003, Harmonization options for EMF standards: proposals of Russian national committee on non- ionizing radiation protection (RNCNIRP). In: *Proceedings of the 3rd International EMF Seminar in China: EMF and Biological Effects. Guilin, China 13 - 17 October, 2003*, pp. 55.

[6] Blackman C, Blank M, Kundi M, Sage C, Carpenter D, Davanipour Z, Gee D, Hardell L, Johansson O, Lai H, Hansson Mild K, Sobel E, Xu X, Chen G and Sage S 2007, *The BioInitiative Report,* http://bioinitiative.org/report/docs/report.pdf

[7] Grigoriev Y 2008, Tenth anniversary of the Russian National Committee on Non-Ionizing Radiation Protection. Annual Proceedings, Russian National Committee on Non-Ionizing Radiation Protection (RNCNIRP), Moscow, 2008, pp. 7-12 – *in Russian.*

[8] SanPiN 2003, Hygienic requirements for the deployment and operation of land mobile radio. SanPiN (Sanitary Provisions and Ecological Norms) guidelines SanPiN 2.1.8/2.2.4.1190-03, Ministry of Health of the Russian Federation / Russian Ministry of Health Protection. http://apekspro.ru/files/SanPiN%202.1.8_2.2.4.1190-03.pdf – *in Russian.*

[9] WHO 2003, Healthy environments for children: WHO backgrounder No. 3, April 2003 http://www.who.int/world-health-day/previous/2003/backgrounder/en/index.html

[10] The Stewart Report 2000 Mobile Phones and Health, *Independent Expert Group on Mobile Phones*, http://www.iegmp.org.uk/report/text.htm

[11] National Academy of Sciences National Research Council 2002, An Assessment of Non-Lethal Weapons Science and Technology by the Naval Studies Board, Division of Engineering and Physical Sciences. National Academies Press.

[12] Vinogradov G I and Dumansky Y D 1974, Changes in the antigenic properties of tissues and autoallergic processes under the influence of MW-energy. *The Bulletin of Experimental Biology and Medicine*, **8**, pp. 76-79 - *in Russian.*

[13] Vinogradov G I and Dumansky Y D 1975, On the sensitization action of ultrahigh frequency electromagnetic fields. *Hygiene and Sanitation,* **9**, pp. 31-35 - *in Russian.*

[14] Shandala M G and Vinogradov G I 1982, Autoallergic effects of electromagnetic energy of the MW-range exposure and their influence on a foetus and posterity. *The Bulletin of the Academy of Medical Sciences of the USSR,* M Medicina Publisher, **10**, pp. 13-16 - *in Russian.*

[15] MMF 2008, Mobile Phone Safety and Use by Children, Mobile Manufacturers Forum (MMF) http://www.mmfai.info/public/docs/eng/MP%20and%20Children.pdf.

IOP Conf. Series: Earth and Environmental Science **10** (2010) 012004 doi:10.1088/1755-1315/10/1/012004

Human Health Effects of EMFs: The Cost of Doing Nothing

David O. Carpenter, MD

Institute for Health and the Environment, University at Albany, Rensselaer, NY 12144-3456 USA

E-Mail: carpent@uamail.albany.edu

Abstract. Everyone is exposed to electromagnetic fields (EMFs) from electricity (extremely low frequency, ELF), communication frequencies and wireless devices (radiofrequency, RF), as well as naturally occurring EMFs. Concern of health hazards from EMFs has increased as the use of mobile phones and other wireless devices has grown in all segments of the population, especially children. While there has been strong evidence for an association between leukemia and residential or occupational exposure to ELF EMFs for many years, the standards in existence are not sufficiently stringent to protect from an increased risk of cancer. ELF EMFs also increase risk of at least two types of neurodegenerative diseases. For RF EMFs, standards are set at levels designed to avoid tissue heating, in spite of many reports of biological effects at intensities too low to cause significant heating. Recent evidence demonstrates elevations in risk of brain cancer and acoustic neuroma only on the side of the head where individuals used their mobile phone. Individuals who begin exposure at younger ages are more vulnerable. These data indicate that the existing standards for radiofrequency exposure are not adequate. While there are many unanswered questions, the cost of doing nothing may result in an increasing number of people, many of them young, developing these diseases.

1. Introduction

It has been known for many years that high energy EMFs (X-rays, gamma rays, cosmic rays) have sufficient energy to directly break chemical bonds, causing damage to molecules ranging from water to DNA that results in cancer and birth defects [1]. Thus these forms of EMF are "ionizing". There is less consensus as to whether lower energy forms of EMFs, such as radiofrequency and ELF EMFs, can cause disease. In spite of strong evidence for such relationships from the biomedical community, most national and international bodies have discounted this evidence, based on the belief that lower energy EMFs cannot possibly cause serious disease [2]. This particular point of view is held by many in the physics and engineering communities, individuals not known for their detailed knowledge of medicine. There are legitimate concerns as to what mechanisms might explain these relationships. The purpose of this review is to provide an overview of the issues, explore both the associations between exposure and disease and the mechanisms that might explain them, and to propose biologically-based standards of exposure which, although difficult to achieve, would be more protective of human health.

While there are a variety of diseases of possible concern, this review will focus on only two major classes, cancer and neurodegenerative disease. This is for two reasons. The evidence for an association with EMF is strongest for these diseases, and these are very serious diseases that cause significant morbidity and mortality in humans.

2. Health Effects of ELF EMFs

There has been evidence that residential exposure to elevated residential magnetic fields results in an increased risk for leukemia since the pioneering studies of Wertheimer and Leeper [3]. Most subsequent studies have confirmed elevated risks of leukemia [4-6], and several meta-analyses have shown significantly elevated odds ratios (ORs) whether exposure was determined through use of wire codes or measured magnetic fields [7-9]. In addition there is evidence that leukemia is elevated in adults employed in occupations that involve elevated exposure to EMFs from electricity [10]. Meta-analyses of occupational exposure have also reported elevated risks for leukemia, with less strong evidence for relations to other kinds of cancer [11]. There has also been a meta-analysis reporting a significant elevation in rates of brain cancer among adults working in "electrical" occupations [12].

There is also strong evidence for a relationship between occupational exposure to EMFs and neurodegenerative diseases, Alzheimer's and amyotrophic lateral sclerosis (ALS). In a meta-analysis of EMFs and Alzheimer's disease, Garcia et al. [13] report an OR ratio of 2.03 (95% CL = 1.38-3.00) for case-control studies, and 1.62 (95% CL = 1.16-2.80) for cohort studies. Ahlbom [14] reviewed seven studies of EMF exposure and ALS, and found an overall significant OR of 1.5 (95% CL = 1.2-1.7). More recent studies of Hakansson et al. [15] found an OR of 2.2 (95% CL = 1.0-4.7) among welders and other workers exposed to elevated magnetic fields.

3. Human Disease from Exposure to RF EMFs

Until recently there has been relatively little attention to RF exposures and human health. Older studies have reported elevations in both leukemia and brain tumors among individuals with occupational exposures to RF (see [16] for references), but results were not very consistent across studies. Recent reports have found elevated rates of leukemia among children who live near AM radio transmitter sites [17-19]. This is the same cancer elevated with exposure to powerline frequency EMFs, suggesting that leukemia is the cancer most likely to show elevated risk with whole body exposure to a variety of EMFs frequencies.

With the advent of enormous increases in the use of mobile phones, we now have a situation in which a very large segment of society is regularly exposed to high levels of RF. In addition, the whole population has increased exposure through the placement of mobile phone towers, wireless buildings and even wireless cities. The strongest evidence for hazards has come from Europe, especially Scandinavia, where mobile phones were initially manufactured, and have been in wide use for a longer period of time as compared to other parts of the world.

Recent studies have found an elevated risk of brain tumors and acoustic neurons in individuals who have used mobile phones regularly for ten years or longer. A meta-analysis by Hardell et al. [20] based on four studies finds an OR of 2.0 (95% CL = 1.2-3.4) for glioma among individuals who have used a mobile phone for ten years or more, but only on the side of the head where the phone was used. There was also an OR of 2.4 (95%CL = 1.1-5.3) for acoustic neuroma among long-term users. Risks for meningioma were elevated, but not statistically significant. Kundi [21] has reported on 33 epidemiological studies, and finds that the combined ORs from these studies show an OR of 1.5 (95% CL = 1.2-1.8) for glioma. There was also a nonsignificant elevation in ORs for acoustic neuroma but no relationship with meningioma. Hopefully, additional information will come from the pooled results of the INTERPHONE study, a 13-nation investigation coordinated by the World Health Organization, which should be available in the near future. Interestingly, the Israeli component of this study has found an elevated risk of parotid gland cancer with long-term mobile phone use [22], but results from the full INTERPHONE study of parotid gland cancers are also not yet available.

There is a particular concern about risks to children exposed to RF. Hardell et al. [23] studied relative risk based on the age when a person began to use a mobile phone, and found that individuals whose use began while they were in their 20's displayed ORs that were higher than those of older persons for both analog and cordless phones when assessed at either >1 or >5 year latency. Later Hardell reported at a meeting (personal communication) that children who began use of a mobile phone prior to the age of 20 had an OR of developing glioma of 5.2 (95% CL = 2.2-12) after only one plus year of mobile phone use, while for all ages the OR was 1.4 (95% CL = 1.1-1.7). The same relative relationship was seen with use of a cordless phone, where use before the age of 20 years gave an OR of 4.4 (95% CL = 1.9-10), whereas for all ages the OR was 1.4 (95% CL = 1.1-1.8). These studies support the conclusion that cordless phones increase both exposure levels and disease by about the same magnitude as do mobile phones, and that use of either results in an increased risk of gliomas.

4. Why Have These Results not been Reflected in New Standards of Exposure?

In spite of this consistency in observations relating to ELF EMFs and leukemia, and the developing evidence for a relationship between mobile phone use and elevated risk of brain cancer and acoustic neuroma, there has been a general failure of governments and international advisory bodies to accept the reported relationships as being cause and effect, and to follow through with standards designed to reduce exposure. This is a consequence of two major scientific problems, as well as the public excitement and support of wireless technologies and the political power of the industry. Animal studies have not consistently demonstrated cancer as a result of exposure to ELF EMFs. In addition, no single mechanism has been identified to be the basis for the development of cancer following exposure to EMFs.

In spite of the widespread belief that most cancer is genetic, recent studies of identical twins have convincingly shown that most cancers result from some environmental exposure and are not primarily genetic [24]. This is not to say that genetics is unimportant, since genetic susceptibility will determine whether or not environmentally-induced cancers develop and become life-threatening. There is also the widespread and mistaken belief that all carcinogens act by causing direct DNA damage, as is the case with ionizing radiation. However, many proven human carcinogens do not cause direct DNA damage. These are identified as "non-mutagenic carcinogens" by the USEPA, and include such well-documented carcinogens as arsenic [25-26] and dioxins [27]. Exact mechanisms are not known to explain the carcinogenicity of either, although a number of possible factors are known. Thus the fact that ELF and RF EMFs are "non-ionizing" does not mean they are not carcinogens. Both ELF and RF EMFs are known to cause gene induction, generate reactive oxygen species, trigger formation of heat shock proteins and cause other alterations in cellular function, any one of which might lead to cancer (see [16] for references and detailed discussion).

The causes of Alzheimer's disease and ALS are also not known. In the case of Alzheimer's disease there is the accumulation in the brain of two different deposits that come from normal proteins, called amyloid plaques and neurofibulary tangles [28]. There is uncertainty whether these deposits cause the disease or are only associated with the disease. Recent evidence suggests that reactive oxygen species formation may be a major factor in causing cell death in Alzheimer's disease [29]. ALS is even less well understood, but the disease is the result of death of upper and lower motoneurons [30]. Both reactive oxygen species [31] and heat-shock proteins [32] have been implicated in the resulting cell death. As with cancer, genetics plays a minor role in etiology of these diseases. There is no reason to discount an association with exposure to EMFs on the basis of lack of firm knowledge of a mechanism, as the mechanisms causing both diseases are uncertain. It is worth noting that reactive oxygen species and heat-shock proteins are implicated in both non-mutagenic carcinogen actions and these neurodegenerative diseases, and that both are altered by EMFs.

5. Proposed EMF Standards that are Based on Studies of Human Health after Exposure

The Bioinitiative Report [33] presents recommendations for standards of EMF exposure that are based on the epidemiological evidence in human populations. For ELF EMFs the proposed standard is 1 mG (0.1 μT), to be compared with the current International Commission on Non-ionizing Radiation Protection (ICNIRP) standard of 1,000 mG (100 μT). For RF radiation the proposed standard is 0.1

μW/cm^2, to be compared with the US Federal Communications Commission standard of 583 μW/cm^2 for 875 MHz cell phone frequency, and 1,000 μW/cm^2 in the frequency range of 1,800 - 1,950 MHz. The difference between these numbers show the magnitude of the problem. There is no question that a sudden imposition of standards so drastically different from those existing would impose severe hardship. However, there is also no question that the human studies clearly indicate that the existing standards are not protective of human health.

The benefits to society derived from electricity and wireless communications are significant, and certainly none of us is willing to return to the pre-electric age. However it is imperative that society at least acknowledge the disparities between current standards and current evidence of adverse health effects. Rigid and sudden imposition of the standards we propose is certainly unrealistic at the present time, but these levels are appropriate goals that could at least be approached by a combination of development of new technology and voluntary changes in behaviors.

6. The Costs of Being Wrong

At present we do not know precisely to what degree risk of cancer and neurodegenerative diseases is increased by excessive exposure to EMFs. Human studies are difficult under any circumstances, but those difficulties are even greater when studying the effects of EMFs. Levels of exposure for each of us vary over the course of every day as we move through our environment and use appliances and mobile phones. This makes exposure assessment extremely difficult. Under the circumstances of poor exposure assessment there is a great likelihood that the total risk is underappreciated.

There is considerable evidence that the developing organism is more vulnerable to several environmental insults than are adults [34-35]. The reality is that children throughout the world are using mobile phones at increasing rates and for long durations. Therefore, if there are real risks, and especially if children are more susceptible, we may be facing an epidemic of brain and other cancers. The concern is increased because, to date, there are few warnings to parents and physicians advising restrictions on use of mobile phones by children. While the evidence at present is certainly less than total proof of a relationship between exposure and elevated risk of cancer, it is sufficiently strong so as to demand precaution. The alternative may be significant increases in certain cancers, especially leukemia and brain cancer. It is not clear whether other kinds of cancer are also at increased risk following exposure, since there has not been much study of, for example, the possible health hazards of wearing a cell phone on your belt and pelvic cancers.

Fortunately, the rates of leukemia and brain cancer are not high. There have recently been significant improvements in treatment of leukemia, especially among children. Kundi [21] has hypothesized that use of mobile phones may increase rates of some forms of brain cancer, such as gliomas, by as much as 50%. Even if this is true, this certainly does not mean that every exposed person will develop brain cancer. Such an increase in brain cancer would still have a significant impact, not only on the individuals affected but also on society, especially given that much of this increase is likely to occur among young people.

Application of the Precautionary Principle is appropriate under the circumstances where there is a demonstrated elevation in rates of serious diseases in humans following elevated EMF exposure, but has many unanswered questions as to mechanisms responsible. We need additional research, of course, and much better exposure assessment. The evidence that we have at present is too convincing to be ignored. Our national and international standards are obsolete, and ignore evidence reported by many different investigators. The lack of certainty with regard to mechanisms and animal models is no reason to ignore studies of human health. Similar lack of certainty regarding mechanisms also exists for some chemicals, yet precautionary measures are commonly taken to reduce exposure. We need the electric and communications industries to be proactive in developing products that can be used with reduced exposures. We need governments and international organizations to set standards that are based on the evidence of whether there are hazards to humans, not on a hypothesis that is not

IOP Conf. Series: Earth and Environmental Science **10** (2010) 012004 doi:10.1088/1755-1315/10/1/012004

credible based on the evidence from animal and cellular studies. Most importantly, we need individuals to understand that personal decisions will significantly impact the level to which they are exposed to both ELF and RF EMFs.

References

[1] Little M P 2003, Risks associated with ionizing radiation. *British Medical Bulletin* **69** pp. 259-275.

[2] Adair R K 2000, Static and low-frequency magnetic field effects: health risk and therapies. *Reports on Progress in Physics* **63** pp. 415-454.

[3] Wertheimer N and Leeper E 1979, Electrical wiring configurations and childhood cancer. *American Journal of Epidemiology* **109** pp. 273-284.

[4] Savitz D A, Wachtel H, Barnes F A, John E M, and Tvrdik J G 1988, Case-control study of childhood cancer and exposure to 60-Hz magnetic fields. *American Journal of Epidemiology* **128** pp. 21-38.

[5] London S J, Thomas D C, Bowman J D, Sobel E, Cheng T C, and Peters J M 1991, Exposure to residential electric and magnetic fields and risk of childhood leukemia. *American Journal of Epidemiology* **134** pp. 923-937.

[6] Feychting M and Ahlbom A 1993, Magnetic fields and cancer in children residing near Swedish high-voltage power lines. *American Journal of Epidemiology* **138** pp. 467-481.

[7] Wartenberg D 1998, Residential magnetic fields and childhood leukemia: A meta-analysis. *American Journal of Public Health* **88** pp. 1787-1794.

[8] Greenland S, Sheppard A R, Kaune W T, Poole C, Kelsh M A and the Childhood Leukemia-EMF Study Group 2000, A pooled analysis of magnetic fields, wire codes, and childhood leukemia. *Epidemiology* **11** pp. 624-634.

[9] Ahlbom A, Day N, Reychting M, Roman E, Skinner J, Doctery J, Linet M, McBride M, Michaelis J, Olsen J H, Tynes T and Verkasalo P K 2000, A pooled analysis of magnetic fields and childhood leukemia. *British Journal of Cancer* **83** pp. 692-698.

[10] Savitz D A and Calle E E 1987, Leukemia and occupational exposure to electromagnetic fields: Review of epidemiologic studies. *Journal of Occupational and Environmental Medicine* **29** pp. 47-51.

[11] Kheifets L I, Afifi A A, Buffler P A, Zhang Z W and Mastkin C C 1997, Occupational electric and magnetic field exposure and leukemia. *Journal of Occupational and Environmental Medicine* **29** pp. 1075-1091.

[12] Kheifets L I, Afifi A A, Buffler P A and Zhang Z W 1995, Occupational electric and magnetic field exposure and brain cancer: A meta-analysis. *Journal of Occupational and Environmental Medicine* **37** pp. 1327-1340.

[13] Garcia AM, Sisternas A and Perez Hoyos S 2008, Occupational exposure to extremely low frequency electric and magnetic fields and Alzheimer disease: a meta-analysis. *International Journal of Epidemiology* 1-12 doi:10.1093/ije/dym295

[14] Ahlbom A 2001, Neurodegenerative diseases, suicide and depressive symptoms in relation to EMF. *Bioelectromagnetics* **Supplement 5** S132-S143.

[15] Hakansson N, Gustavsson P, Johansen C and Floderus B 2003, Neurodegenerative diseases in welders and other workers exposed to high levels of magnetic fields. *Epidemiology* **14** pp. 413-419.

[16] BioInitiative: Biologically-based Exposure Standards for Low-Intensity Electromagnetic Radiation, http://www.bioinitiative.org/

[17] Michelozzi P, Capon A, Kirchmayer U, Forastiere F, Biggeri A, Barca A and Perucci C A 2002, Adult and childhood leukemia near a high-power radio station in Rome, Italy. *American Journal of Epidemiology* **155** pp. 1096-1103.

[18] Park S K, Ha M and Im H J 2004, Ecological study on residences in the vicinity of AM radio broadcasting towers and cancer death: preliminary observations in Korea. *International Archives of Occupational and Environmental Health* **77** pp. 387-394.

[19] Ha M, Im H, Lee M, Kim H J, Kim B-C, Gimm Y-M and Pak J-K 2007, Radio-frequency radiation exposure from AM radio transmitters and childhood leukemia and brain cancer. *American Journal of Epidemiology* **166** pp. 270-279.
[20] Hardell L, Carlberg M, Soderqvist F and Mild K H 2008, Meta-analysis of long-term mobile phone use and the association with brain tumours. *International Journal of Oncology* **32** pp. 1097-1103.
[21] Kundi M 2008, The controversy about a possible relationship between mobile phone use and cancer. *Environmental Health Perspectives* **117**, pp. 316–324. doi:10.1289/ehp.11902
[22] Sadetzki S, Chetrit A, Jarus-Hakak A, Cardis E, Deutch Y, Duvdevani S, Zultan A, Novikov I, Freedman L and Wolf M 2008, Cellular phone use and risk of benign and malignant parotid gland tumors. A nationwide case-control study. *American Journal of Epidemiology* **167** pp. 457-467.
[23] Hardell L, Mild K H, Carlberg M and Hallquist A 2004, Cellular and cordless telephone use and the association with brain tumors in different age groups. *Archives of Environmental Health* **59** pp. 653-659.
[24] Lichtenstein P, Holm N V, Verkasalo P K, Iliadou A, Kaprio J, Koskenvuo M, Pukkala E, Skytthe A and Hemminki K 2000, Environmental and heritable factors in the causation of cancer. Analyses of cohorts of twins from Sweden, Denmark and Finland. *The New England Journal of Medicine* **343** pp. 78-85.
[25] Gebel T 2000, Confounding variables in the environmental toxicology of arsenic *Toxicology* **144** pp. 155-162.
[26] Hughes M F 2002, Arsenic toxicity and potential mechanisms of action. *Toxicology Letters* **133** pp. 1-16.
[27] Nowak R 1991, Dioxin: Environmental toxin, molecular probe. *Journal of NIH Research* **3** pp. 49-53.
[28] Vickers J C, Dickson T C, Adlard P A, Saunders L H, King C E and McCormack G 2000, The cause of neuronal degeneration in Alzheimer's disease. *Progress in Neurobiology* **60** pp. 139-165.
[29] Tabner B J, Turnbull S, El-Agnaf O M A and Allsop D 2002, Formation of hydrogen peroxide and hydroxyl radicals from Aβ and α-synuclein as a possible mechanism of cell death in Alzheimer's disease and Parkinson's disease *Free Radical Biology & Medicine* **32** pp. 1076-1088.
[30] Wong P C, Rothstein J D and Price D L 1998, The genetic and molecular mechanisms of motor neuron disease. *Current Opinion in Neurobiology* **8** pp. 791-799.
[31] Liu R, Althaus J S, Ellerbrock B R, Becker D A and Gurney M E 1998, Enhanced oxygen radical production in a transgenic mouse model of familial amyotrophic lateral sclerosis. *Annals of Neurology* **44** pp. 763-770.
[32] Okado-Matsumoto A and Fridovich I 2002, Amyotrophic lateral sclerosis: A proposed mechanism. *Proceedings of the National Academy of Sciences* **99** pp. 9010-9014.
[33] Blackman C, Blank M, Kundi M, Sage C, Carpenter D, Davanipour Z, Gee D, Hardell L, Johansson O, Lai H, Hansson Mild K, Sobel E, Xu X, Chen G and Sage S 2007, *The BioInitiative Report,* http://bioinitiative.org/report/docs/report.pdf
[34] Ginsberg G L 2003, Assessing cancer risks from short-term exposures in children. Risk Analysis **23** pp. 19-34.
[35] Landrigan P J and Garg A 2005, Children are not little adults. *Children's Health and the Environment. A Global Perspective* Pronczuk-Garbino J (Ed), World Health Organization, Geneva, Chapter 1, pp. 3-16.

IOP Conf. Series: Earth and Environmental Science **10** (2010) 012005 doi:10.1088/1755-1315/10/1/012005

Aspects of Studies on the Functional Impairment Electrohypersensitivity

Olle Johansson

The Experimental Dermatology Unit, Department of Neuroscience, Karolinska Institute, 171 77 Stockholm, Sweden and The Royal Institute of Technology, 100 44 Stockholm, Sweden.

Email: olle.johansson@ki.se

Abstract. Persons, claiming to suffer from exposure to electromagnetic fields, have been described in the literature. In Sweden, electrohypersensitivity (EHS) is an officially fully recognized functional impairment (i.e., it is not regarded as a disease). Survey studies show that somewhere between 230,000– 290,000 Swedish men and women – out of a population of 9,000,000 - report a variety of symptoms when being in contact with electromagnetic field (EMF) sources. Swedish electrohypersensitive people have their own handicap organization, The Swedish Association for the Electrohypersensitive, which has its own website in both Swedish and English. This organization is included in the Swedish Disability Federation (Handikappförbundens SamarbetsOrgan; HSO). One aim of our studies has been to investigate possible alterations, in the cellular and neuronal systems of these persons' skin. In summary, it is evident from our preliminary data that various alterations are present in the electrohypersensitive persons' skin that are not indicated in the skin of normal healthy volunteers.

1. Introduction

In Sweden, electrohypersensitivity (EHS) is now an officially fully recognized functional impairment (i.e., it is not regarded as a disease). Survey studies [1] show that somewhere between 230,000–290,000 Swedish men and women – out of a population of 9,000,000 - report a variety of symptoms when being in contact with manmade sources of electromagnetic fields (EMF) [1].

2. Rights for Electrohypersensitive Individuals in Sweden

Swedish electrohypersensitive people have their own handicap organization, The Swedish Association for the Electrohypersensitive [2], which has its own website, http://www.feb.se, in both Swedish and English. This organization is included in the Swedish Disability Federation (Handikappförbundens SamarbetsOrgan; HSO). HSO is the united voice of the Swedish disability associations towards the government, the parliament, and national authorities, and is a cooperative body that today consists of 43 national disability organisations (with The Swedish Association for the Electrohypersensitive being 1 of these 43 organisations) comprised of about 500,000 individual members. It has its own website in Swedish, parts of which are also in English [3].

Swedish municipalities, of course, have to follow the UN 22 "Standard Rules on the equalisation of opportunities for people with disabilities" ("Standardregler för att tillförsäkra människor med funktionsnedsättning delaktighet och jämlikhet" [4]. Since 2007 they have been upgraded into the UN "Convention on Human Rights for Persons with Functional Impairments") [4]. As a result of this, all people with disabilities shall, thus, be given the assistance and service they have the right to according to the Swedish Act concerning Support and Service for Persons with Certain Functional Impairments ("LSS-lagen") and the Swedish Social Services Act ("Socialtjänstlagen").

People with disabilities, thus, now have many different rights and can get different kinds of support. The purpose of those rights and the support is to give every person the chance to live like everyone else. It is intended that as a result of this everyone who lives in the Swedish municipalities shall be able to lead a normal life and the municipalities must have correct knowledge and be able to reach the people who need support and service. People with disabilities shall be able to get extra support so that they can live, work, study, or do things they enjoy in their free time. The municipalities are responsible for making sure that everyone gets enough support. Everyone is required to show respect and remember that such men and women may need individual and different kinds of support.

3. The Environment, Impairments and the Individual

In Sweden, impairments are viewed from the point of the environment. No human being is in itself impaired, there are instead shortcomings in the environment that cause the impairment (as with the lack of ramps for the person in a wheelchair or rooms requiring electrosanitisation for the person with EHS). This environment-related impairment view, furthermore, means that even though one does not have a complete scientifically based complete explanation for the impairment EHS, and in contrast to what many individuals involved in EMF discourse at present, the person with EHS shall always be met in a respectful way and with all necessary support with the goal to eliminate the impairment. This implies that the person with EHS shall have the opportunity to live and work in an electrosanitised environment.

This view can fully be motivated in relation to the present national and international handicap laws and regulations, including the UN 22 Standard Rules/UN Convention and the Swedish action plan for persons with impairments [5]. Also, the Human Rights Act in the EU fully applies.

4. Implications of Electrohypersensitivity

Many people who are electrically hypersensitive in Sweden, as well as elsewhere, have a living hell today. They may be dismissed from their jobs, refused housing adaptation, denied sick leave, compelled to move, etc. While this is happening, a lot of people are talking about the necessity of taking EHS seriously. This talk, however, is not generally leading to anything since it is action that is required. In the first place, working and housing environments must be created where people who are electrohypersensitive can live and work.

The design of our physical, intellectual and social environment is not something laid down in the constitution or divinely ordained. It is not a law of nature that everyone has to have mobile telephones, to use high-frequency lighting apparatus and wireless Internet. However, if someone might say, we have "diversity" in Sweden – surely that applies to everyone? We may all listen to our favourite music, dress according to taste, and vote for the party of our choice. But, unfortunately, when it comes to disabilities, adaptation and assimilation generally still take priority over the minority's right to access – despite the electrically hypersensitive, and others with disabilities, having other requirements that need addressed.

This interdisciplinary issue with regard to those with disabilities applies internationally right across the scientific divide, the healthcare sector and social services, and into user-focused lifestyle and design initiatives. There is a need to educate and design for all to ensure that consumer products, the built environment and all its components (as designed and specified by architects, engineers and others) to positively allow for the inclusion of those with disabilities. As part of this it is also suggested that current building standards, regulations and guidance documents should be developed to incorporate guidelines on how to create suitable environments for those with EHS, as are now generally given for a number of other disabilities.

It is necessary to encourage the necessity and benefits of good EHS research to policymakers and the developers of new technologies to address the major concerns that presently exist. It is also necessary to adequately fund more multidisciplinary work in this field to allow for the research and development of more 'user-friendly' environments, particularly as related to biological effects.

5. Cellular and Neuronal Studies of Electrohypersensitivity

We have investigated the presence of intraepidermal nerve fibers in normal human skin from healthy volunteers using the new marker PGP 9.5 [6-8]. The intraepidermal nerve fibers are found as close as 20–40 μm from the surface, which makes it highly possible that weak electromagnetic fields may affect them.

In facial skin samples of electrohypersensitive persons, the most common finding is a profound increase of mast cells. Nowadays we do not only use histamine, but also other mast cell markers such as chymase and tryptase, but the pattern is still the same as reported previously for other electrohypersensitive persons [9]. From these studies, it is clear that the number of mast cells in the upper dermis is increased in the EHS group. A different pattern of mast cell distribution also occurred in the EHS group, namely, the normally empty zone between the dermo-epidermal junction and mid-to-upper dermis disappeared in the EHS group and, instead, this zone had a high density of mast cell infiltration. These cells also seemed to have a tendency to migrate towards the epidermis (=epidermiotrophism) and many of them emptied their granular content (=degranulation) in the dermal papillary layer. Furthermore, more degranulated mast cells could be seen in the dermal reticular layer in the EHS group, especially in those cases which had the mast cell epidermiotrophism phenomenon described above. Finally, in the EHS group, the cytoplasmic granules were more densely distributed and more strongly stained than in the control group, and, generally, the size of the infiltrating mast cells was found to be larger in the EHS group as well. It should be noted that increases of similar nature later on were demonstrated in an experimental situation employing normal healthy volunteers in front of cathode ray tube (CRT) monitors, including ordinary household television sets [10].

In one of the early papers [11], we made a sensational finding when we exposed two electrically sensitive individuals to a TV monitor situated at a distance of 40-50 cm away from them. When we looked at their skin under a microscope, we found something that surprised us. In this article, we used an open-field provocation, in front of an ordinary TV set, of persons regarding themselves as suffering from skin problems due to work at video display terminals. Employing immunohistochemistry, in combination with a wide range of antisera directed towards cellular and neurochemical markers, we were able to show a high-to-very high number of somatostatin-immunoreactive dendritic cells as well as histamine-positive mast cells in skin biopsies from the anterior neck taken before the start of the provocation. At the end of the provocation the number of mast cells was unchanged, however, the somatostatin-positive cells had seemingly disappeared. The reason for this latter finding is discussed in terms of loss of immunoreactivity, increase of breakdown, etc. The high number of mast cells present may explain the clinical symptoms of itch, pain, edema, and erythema.

We have compared facial skin from electrohypersensitive individuals with corresponding material from normal healthy volunteers [12]. The aim of the study was to evaluate possible markers to be used for future double-blind or blind provocation investigations. Differences were found for the biological markers calcitonin gene-related peptide (CGRP), somatostatin (SOM), vasoactive intestinal polypeptide (VIP), peptide histidine isoleucine amide (PHI), neuropeptide tyrosine (NPY), protein S-100 (S-100), neuron-specific enolase (NSE), protein gene product (PGP) 9.5, and phenylethanolamine N-methyltransferase (PNMT). The overall impression in the blind-coded material was such that it turned out easy to blindly separate the two groups from each other. However, no single marker was 100% able to pin-point the difference, although some were quite powerful in doing so (CGRP, SOM, S-100). In our ongoing investigations, we have also found alterations of the Merkel cell number in the facial skin of electrohypersensitive persons [13]. However, it has to be pointed out that we cannot, based upon those results, draw any definitive conclusions about the cause of the changes observed. Blind or double-blind provocations in a controlled environment [10] are necessary to elucidate the underlying causes for the changes reported in this particular investigation.

6. Discussion

As previously mentioned Sweden has officially recognised EHS as a disability. However, a lot of work still has to be carried out by the electrically hypersensitive, as well as for them, and their disability organisation, the Swedish Association for the Electrohypersensitive (FEB) to achieve complete equality. Accessibility and adaptation are key issues for allowing EHS, and others with recognized functional impairments, to gain/regain their rightful independence. Such support can also benefit society in general.

Society must recognise in practical applications the right of the electrically hypersensitive to be different, to their distinguishing feature. Society must recognise the right of the electrically hypersensitive to have an equal life in a society based on equality. Treating members of the community equally is not something that should be done as a favour; nor is it something that any parliament or government should politely request other inhabitants to provide others with. Equality is not something to be done "out of the goodness of one's heart". It is something one does because it is expected of every citizen, because inaccessibility and discrimination are prohibited by law.

Some doctors and dentists have described at an early stage the electrically hypersensitive as "old crones in the throes of the menopause", "the poorly educated", "hypochondriacs", "radiation ladies", etc. These prejudiced care-providers used these terms despite often never having met an electrically hypersensitive person or carrying out research in the field.

At times, the view taken of the electrically hypersensitive resembled a medical pogrom with frightening overtones. Even more frightening is that the situation today, well over two decades later, is allowed to be much the same – and not only in Sweden. The entire public debate has been characterised by various degrees of acceptance and dismissal, grand words, commitments and far-reaching promises, which have often been shown to lack the necessary effect. This nonchalance and lack of commitment has per se been particularly insulting to those that are afflicted, particularly when there are often simple measures, such as not using wifi in buildings, which can be readily undertaken that could often greatly help ease the situation [14]. Is all this inaction something that we should be proud about? How much simpler, cheaper and more creditable it would have been to have provided the electrically hypersensitive immediately with the assistance and solidarity that we like to boast about and seek to create more biologically friendly environments? An immediate decision on a housing adaptation grant, disability allowance, etc., would have been self-evident as would increased dialogue with those creating electronic devices. Now we have instead endured almost three decades of hostilities and denial with big losers, namely the electrically hypersensitive and their next-of-kin.

There is much that could be done: educating architects, planners, scientists, technologists and the general public more effectively about EHS, its causes and how it can be minimized; undertaking properly funded independent multidisciplinary research into EHS and showing that such work can make a difference. Creating work, home and general environments that are more user-friendly for EHS sufferers so that they feel included and not excluded in the rich tapestry of life.

6.1 The Need to Empower the Individual

It is proposed that, as with those with other recognized disabilities, the electrically hypersensitive must therefore, in every situation and by all available means, demand respect, representation and power. They shall very clearly reject all approaches which reflect a mentality of "feeling pity for them" or "caring for them". Inaccessibility is not a personal problem. It is a problem for society. Inaccessibility is not about attitudes. It is about discrimination. And discriminatory actions and conduct shall not be dealt with by well-meaning talk about treatment. Discrimination is already illegal!

6.2 Disability Plan Policies

The former Swedish Prime Minister Göran Persson has declared that 2010 should be the final target year for the "National Disability Plan Policy" – "From patient to citizen" – adopted in 2000, according to which the whole of Sweden is to be completely adapted to those with disabilities [5]. In addition, there is the EU "Human Rights Act" and the UN "Standard Rules on Equalization of Opportunities for People with Disabilities", nowadays updated to the UN "Convention on Human Rights for Persons with Functional Impairments" [4].

One of the most important ideas in these documents is the "principle of accessibility" stating that people with disabilities are to have full access to public services. This is the basis for the fast introduction in Sweden and other countries of kneeling buses, wheelchair ramps, hearing loops, automatic door-openers, beveled pavement edges, etc. However, it has to be asked where are the measures for the electrically hypersensitive?

How are they to be able to be a normal part of the community with complete access to council offices, post offices, means of transport, cinemas, restaurants, hospital care and other facilities when electromagnetic pollution can affect them detrimentally? Providing every electrically hypersensitive person with individually designed assistance, good care and stimulation to create participation in the community are very responsible tasks that require a high level of skill.

Considerably greater demands must be made on education, training and work supervision than has been the case to date. Sharing experiences between different activities must take place much more smoothly and in a way that is completely free of prestige.

6.3 Future Research

The task of medical research must be to increase knowledge about different disabilities, as is in fact clearly defined in the Karolinska Institute's policy declaration, among other places. The over 200,000 people with the disability EHS in Sweden, as well as all the millions around the world, can only be provided with the right tools by our acquiring knowledge about their disability and immediately putting this knowledge to use to adapt accessibility and develop safer technology. Examples of the organisations that have responded to this challenge are TCO Development (Swedish Confederation for Professional Employees), Sif (Swedish Union of Clerical and Technical Employees in Industry) and FEB. But are all those who can make the necessary changes prepared to commit to such measures?

I have said it many times: It is a true must that fully financed and truly independent research projects should be initiated immediately to ascertain the effects of EMF on public health particularly on vulnerable sectors of the population, and help allay concerns over potential perceived biases in findings. Ideally, the participating scientists *would be* completely devoid of commercial interests of any sort to further reinforce the integrity of their findings and their neutrality. Of course, the determination of such matters are the responsibility of each elected government in each country, and is of special importance for people with the functional impairment electrohypersensitivity (EHS).

We must, here and now, decide how we shall build human bridges of knowledge, for persons with EHS, and others with recognized functional impairments, as well as for all future groups asking and begging for our professional assistance. Should science be perceived of in the form of problem-solving research, or rather a playground for careerists and big industry? Is it the general view of this conference held at the Institute of Physics, as well as the one, "EMF and Health – A Global Issue --- Exploring appropriate precautionary approaches", held just before, September 8-9, 2008, at the Royal Society, London, U.K., that we only shall build half- or three-quarter-ready bridges of knowledge...? Or shall we rather build complete bridges…? It is a truly international issue that we should embrace together.

As President Obama noted when President-elect, "The truth is that promoting science isn't just about promoting resources – it is about protecting free and open inquiry. It's about ensuring that facts and evidence are never twisted or obscured by politics or ideology. It's about listening to what our scientists have to say, even when it's inconvenient – especially when it is inconvenient. Because the highest purpose is the search for knowledge, truth and a greater understanding of the world around us." [15]

7. Conclusion

People with the disability EHS have an exciting future ahead of them. Work has already started to produce results on a number of these issues, including building planning and construction [14]. However, a lot more can and must be done. There are still many years of hard, constructive and consistent work waiting. I hope that everyone with the functional impairment EHS all over the world can find a common platform for this work, and help move the work forward through community and solidarity in the interdisciplinary work that is required.

It is now so important for everyone to decide on the continued direction of their activities as well as their focus and for people to work together with integrity as a team. A continued energetic action plan for the next few months and years together with a broad collaboration with other disability associations are of the utmost importance here. As clearly stated by the UN [4], there must be an end to nonchalance, lack of consideration, indifference and lack of respect on the part of society for those with disabilities for all our sakes. As moral human beings we should never accept negative discriminatory treatment or an insulting special treatment of those with impairments.

Furthermore, another very important issue is the one about how persons with the functional impairment EHS should be met and respected... and how they want themselves to be met and respected? Are they to be viewed as psychiatric patients? As somatic patients? Or as persons with a functional impairment? In Sweden, the latter is true, but will people with a fully accepted impairment in Sweden find it written off as psychological should they travel to the UK, or elsewhere in the world? This would not only be contrary to their human rights, but also a major scandal!

In particular, as levels of individuals with EHS is predicted to increase [16], it appears that this is a problem that must be openly addressed in far greater detail by governments and parliaments, working in conjunction with scientists and those involved in industry and related fields, so that solutions can be found and complete accessability achieved.

Mother Teresa's Nobel Lecture contained a simple message: "the poor must know that we love them". I now say about persons with the functional impairment EHS: "they must know that we respect them, and that we respect and follow the regulations, laws and conventions already decided upon". If it is difficult for you to take a stand on this, just put yourself into the situation of waking up one morning, realizing that you, a family member, a relative, a friend or a colleague has become electrohypersensitive – how would you like to be met? And how do you want to see them? Would you want us to seek the answers to it?

Acknowledgements
Supported by the Karolinska Institute, the Cancer and Allergy Foundation (Cancer- och Allergifonden), Stiftelsen Tornspiran, and The Help Foundation (Hjälpfonden).

References
[1] Miljöhälsorapport 2001, The National Board of Health and Welfare (in Swedish; ISBN: 91-7201-495-4).
[2] The Swedish Association for the Electrohypersensitive (http://www.feb.se; the website has an English version).
[3] HSO, Handikappförbundens Samarbets Organ; http://www.hso.se.
[4] UN 22 Standard Rules, see website: http://www.un.org.
[5] The Swedish action plan for persons with impairments ("Den nationella handlingplanen för handikappolitiken – Från patient till medborgare"; proposition 1999/2000:79, bet. 1999/2000:SoU14).
[6] Hilliges M, Wang L, Johansson O 1995, Ultrastructural evidence for nerve fibers within all vital layers of the human epidermis, *The Journal of Investigative Dermatology*, **104**, pp. 134–137.
[7] Johansson O, Wang L, Hilliges M, Liang Y 1999, Intraepidermal nerves in human skin: PGP 9.5 immunohistochemistry with special reference to the nerve density in skin from different body regions, *Journal of the Peripheral Nervous System* **4**, pp. 43–52.
[8] Wang L, Hilliges M, Jernberg T, Wiegleb-Edstrom D, Johansson O 1990, Protein gene product 9.5-immunoreactive nerve fibres and cells in human skin, *Cell and Tissue Research* **261**, pp. 25–33.
[9] Johansson O, Liu P-Y 1995, "Electrosensitivity", "electrosupersensitivity" and "screen dermatitis": preliminary observations from on-going studies in the human skin, In: Simunic, D., ed. Proceedings of the COST 244: Biomedical Effects of Electromagnetic Fields – Workshop on Electromagnetic Hypersensitivity. Brussels/Graz: EU/EC (DG XIII), pp. 52–57.
[10] Johansson O, Gangi S, Liang Y, Yoshimura K, Jing C, Liu P-Y 2001, Cutaneous mast cells are altered in normal healthy volunteers sitting in front of ordinary TVs/PCs—results from open-field provocation experiments, *Journal of Cutaneous Pathology* **28**, pp. 513–519.
[11] Johansson O, Hilliges M, Björnhagen V, Hall K 1994, Skin changes in patients claiming to suffer from "screen dermatitis": a two-case open-field provocation study, *Experimental Dermatology* **3**, pp. 234–238.
[12] Johansson O, Hilliges M, Han S W 1996, A screening of skin changes, with special emphasis on neurochemical marker antibody evaluation, in patients claiming to suffer from screen dermatitis as compared to normal healthy controls, *Experimental Dermatology* **5**, pp. 279–285.
[13] Yoshimura K, Liang Y, Kobayashi K, Johansson O, Alteration of the Merkel cell number in the facial skin of electrosensitive persons—a morphological study, To be submitted.
[14] Lindberg E-R 2010, Building planning for persons with the functional impairment electrohypersensitivity - a project in the voltage field between belief and knowledge (in Swedish with English summaries), Doctoral Dissertation (in manuscript), The Royal Institute of Technology, School of Architecture and the Built Environment, and the Karolinska Institute, Department of Neuroscience, Stockholm, Sweden.
[15] Obama B 2008, "The search for knowledge, truth and a greater understanding of the world around us", President-elect Obama's Weekly Address 20 December 2008, Change. Gov, The Office of the President Elect, http:://www.change.gov/newsroom/entry/the_search_for_knowledge_truth_and_a_greater_understanding_of_the_world_aro/%5d
[16] Hallberg Ö, Oberfeld G 2006, Letter to the Editor: Will we all become electrosensitive?", *Electromagnetic Biology and Medicine* **25**, pp. 189–191.

[For additional reading and references to the relevant scientific literature, please contact the author by e-mail at olle.johansson@ki.se].

IOP Conf. Series: Earth and Environmental Science **10** (2010) 012006 doi:10.1088/1755-1315/10/1/012006

Medical Applications of Electromagnetic Fields

Henry C. Lai[1] and Narendra P. Singh[1]

[1]Department of Bioengineering, University of Washington, Seattle, WA 98195-5061, USA.

Email: hlai@u.washington.edu

Abstract. In this article, we describe two possible applications of low-intensity non-ionizing electromagnetic fields (EMF) for the treatment of malaria and cancer, respectively. In malaria treatment, a low-intensity extremely-low frequency magnetic field can be used to induce vibration of hemozoin, a super-paramagnetic polymer particle, inside malaria parasites. This disturbance could cause free radical and mechanical damages leading to the death of the parasite. This concept has been tested in vitro on malaria parasites and found to be effective. This may provide a low cost effective treatment for malaria infection in humans. The rationale for cancer treatment using low-intensity EMF is based on two concepts that have been well established in the literature: (1) low-intensity non-thermal EMF enhances cytotoxic free radicals via the iron-mediated Fenton reaction; and (2) cancer cells have higher amounts of free iron, thus are more susceptible to the cytotoxic effects of EMF. Since normal cells contain minimal amount of free iron, the effect would be selectively targeting cancer cells. Thus, no adverse side effect would be expected as in traditional chemotherapy and radiation therapy. This concept has also been tested on human cancer cell and normal cells in vitro and proved to be feasible.

1. Introduction

Non-ionizing electromagnetic fields (EMF), from extremely-low frequency to radiofrequency, have been shown to cause biological effects even at low intensity. Some of these effects may be applied for medical treatments. One example is their effect on the physiology of bone cells that led to the use of these fields for treatment of bone fracture and to improve and facilitate bone healing. In this article, we describe two areas of research on medical applications of electromagnetic fields that we have been engaging in, namely, treatments of malaria infection and cancer. Even through preliminary data support the feasibility of these two applications, mechanisms of effect are still speculative and further research is needed to fully develop them for human treatment.

2. Treatment of Malaria

Malaria affects and kills millions of people in the world, particularly in developing countries. Resistance to anti-malarial drugs is widespread and has been a major problem in malaria treatment and eradication. The need novel approaches for malaria treatment is enormous. We have been investigating such a novel approach for the treatment of malaria using an alternating magnetic field.

The basis of this approach is that malaria parasites generate a super-paramagnetic particle known as hemozoin during their process of feeding. The parasite feeds on hemoglobin in erythrocytes in the blood of the host. A hemoglobin molecule consists of a protein-moiety globin and heme, an iron-containing porphyrin molecule. The globin portion of the hemoglobin molecule is broken down inside the food vacuole of the parasite to amino acids which are used by the parasite for protein synthesis and reproduction [1, 2]. The heme portion, ferriprotoporphyrin IX, is left intact, since the parasite lacks the enzyme to break it down. Free heme is highly toxic to the malaria parasite, because it is an inhibitor of an enzyme (Na^+/K^+-ATPase) that is involved in metabolic energy production [3]. Free heme can also cause a chain reaction of free radical-induced oxidation of unsaturated fatty acids, leading to membrane damage [4]. To eliminate their toxic effects, heme molecules in the food vacuoles of the parasite are packed by an enzymatic process into a polymer called hemozoin. In hemozoin, the heme polymer is formed by covalent bonding of the iron of one heme molecule with the oxygen of the carboxylate group in one of the side chains of another heme molecule [5, 6]. This molecular arrangement makes hemozoin super-paramagnetic [7] and it behaves like a small bar magnet in a magnetic field. We hypothesized that exposure of the parasites to an alternating magnetic field would shake hemozoin molecules inside the parasite. If the energy exerted on the molecules by the magnetic field is large enough, the rate of enzymatic formation of hemozoin would be decreased. Accumulation of free heme in the parasite would then produce toxic effects.

We have tested the effects of alternating magnetic fields on growth of malaria parasites cultured in erythrocytes. Our data show that exposure of malaria parasites cultured in erythrocytes to a low-frequency, low-intensity alternating magnetic field (5 Hz with a duty cycle of 1 second on, 1.5 seconds off and peak-to-peak intensity of 1.5 millitesla) significantly decreased the number of parasites and synthesis of macromolecules (incorporation of [^{3}H]-hypoxanthine into DNA and RNA) by the parasite. Details of the experiment and the results have been published [8]. The 5 Hz frequency was chosen arbitrarily in this preliminary experiment. It is possible that other frequencies may work better.

We propose two mechanisms by which an alternating magnetic field may be effective in causing a deleterious effect on malaria parasites. First, if the field is applied during the erythrocytic stage of the parasite lifecycle in which free heme molecules are being actively polymerized into hemozoin, it may inhibit polymerization of heme into hemozoin. Excess free heme leads to oxidative molecular damages and, in turn, growth retardation and death of the parasite. The second mechanism is that an oscillating magnetic field would cause oscillation of hemozoin, leading to mechanical damage to the organelles of the parasite. Mechanical stress is known to cause cell death via apoptosis [9].

An important advantage of using alternating magnetic fields for the treatment of malaria is that development of resistance would be unlikely. The alternating magnetic field would act directly on hemozoin rather than on an enzyme or other gene product, thus producing no biological selection. Another possibility is that there may be an additive effect, in which alternating magnetic fields would increase the sensitivity of parasites to low concentration of anti-malarial drugs and enhance the effectiveness of treatment.

Further research needed to be carried out to identify the optimal magnetic field exposure conditions to disturb the growth of the malarial parasite. Since the magnetic moment of a hemozoin molecule is proportional to its mass, a resonant frequency probably exists that can make a magnetic field most efficient for malaria treatment.

3. Treatment of Cancer

In previous research, we found that acute (2 hr) exposure to a 60-Hz magnetic field caused DNA single and double strand breaks [10], DNA-protein and DNA-DNA crosslinks [11], and increased apoptosis [12] in brain cells of rats. The effects were mediated by free radicals, since they could be blocked by pretreatment with free radical scavengers [12, 13]. Further studies showed that the effects involved iron because pretreating rats before magnetic field exposure with the iron-chelator deferriprone eliminated the effects [13]. We proposed that magnetic fields generate free radicals via the Fenton reaction (Fig. 1). Other research has also supported the notion that electromagnetic fields in both the extremely-low frequency and radiofrequency ranges, enhance free radical activity/formation in cells and iron could play a role in the process [14].

Iron plays a vital role in cell functions and growth, e.g., in energy metabolism and DNA synthesis. Special molecular mechanisms have evolved for the transport of iron into cells. In vertebrates, an iron transport system involves a specific interaction between the iron-binding protein transferrin in the extracellular fluid and a cell surface transferrin receptors that results in a facilitated transport of iron across cell membrane via receptor-mediated endocytosis [15]. Due to their rapid rate of division, most cancer cells have high rates of iron intake [16] and express a high cell surface concentration of transferrin receptors [17] than normal cells. In general, the aggressiveness of a tumor is positively correlated with cell surface transferrin receptor concentration of its cells. For example, breast cancer cells have 5-15 times of transferrin receptors on their cell surface than normal breast cells [18], and they do take up more iron than normal breast cells [19].

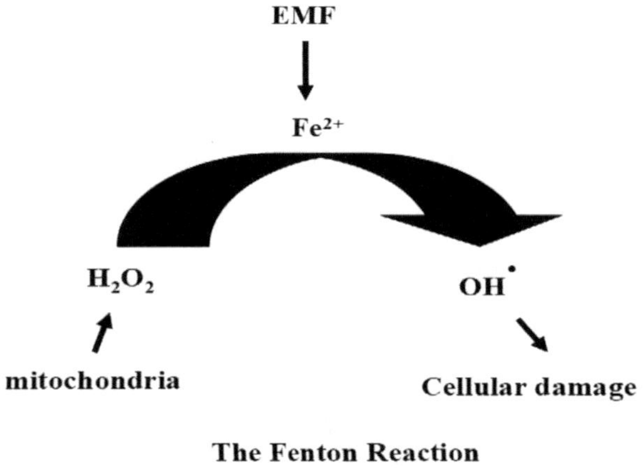

Figure 1. The Fenton reaction: an iron-catalyzed conversion of hydrogen peroxide to a more powerful and damaging hydroxyl free radical that can cause molecular damages and cell death.

Since cellular responses to magnetic fields involve an iron-dependent process, we hypothesize that cancer cells are more responsible to magnetic fields than normal cells. To test this hypothesis, we exposed Molt-4 cells, a human leukaemia cell line, to a 60-Hz magnetic field in the presence and absence of added holotransferrin (i.e., iron loaded transferrin) in the medium. Holotransferrin, as described above, transports iron into cells and increases iron content intracellularly. Molt-4 cells are expected to uptake a large amount of iron when provided with holotransferrin since they have high cell surface concentration of transferrin receptor [20, 21]. An increase in intracellular iron concentration would make these cells more susceptible to the effect of magnetic field. However, since normal lymphocytes do not have high transferrin receptors and do not import a lot of iron via transferrin, they would be less susceptible to the effect of magnetic fields. The following is a description of the experimental procedures.

Molt-4 cells (American Type Culture Collection, Rockville, MD) were grown in 100% humidity at 37°C in 5% CO_2 in air, in RPMI-1640 medium (Life Technologies, Gaithersberg, MD) with 10% fetal bovine serum (Hyclone, Logan, UT). At 24 hrs after splitting into two by adding culture medium, 0.1 ml aliquots were put into microfuge tubes. Half of these samples were incubated with the addition of human holotransferrin (1 mg/ml, Sigma Chemical Co, St. Louis, MO) for 1 hr. Samples from holotransferrin-treated and non-treated cultures were then each further divided into two sets. One of the sets was exposed to a 0.25 mT 60 Hz sinusoidal magnetic field for 2 hrs in an incubator at 37°C. The other was incubated at 37°C without magnetic field exposure. Therefore, there were four treatment conditions: holotransferrin/ magnetic field, holotransferrin/no magnetic field; no holotransferrin/magnetic field, and no holotransferrin/no magnetic field. Exposure to 60-Hz magnetic field was done in a Helmholtz coil exposure system consisting of two 16 cm diameter coils (250 turns/coil) encased in Perspex. Current input to the coils was adjusted using a variac. The flux density (0.25 mT) within the coils was determined by an Enertech Emdex II magnetic field meter. Similarly, 50 μl of human whole blood obtained from a finger prick was mixed with 1 ml of RPMI-1640 and divided into 0.1 ml samples. These samples were subjected to the same procedures of holotransferrin treatment and magnetic field exposure as described above.

Cell counts were made from each cell sample immediately before and after exposure (2 hrs) and at 22 hrs after exposure. After thoroughly suspending cells, 20 μl of cells were mixed well with 20 μl of 10 μg per ml of acridine orange and 10 μl of this was loaded in a hemocytometer chamber. Acridine orange is a DNA intercalating dye and stains DNA and RNA only. This allowed us to count leukocytes in whole blood without isolating them from the blood. Duplicate cell counts were made from each sample.

Cell count of a sample at a certain time point was converted as a ratio of the count of that sample at time zero (i.e., start of exposure). Four experiments each were run with Molt-4 cells and leukocytes and average responses of the experiments were calculated and plotted. Data were then compared by the Mann-Whitney U test and a difference at $p < .05$ was considered statistically significant.

Cell counts of Molt-4 cells after various times after magnetic field exposure were compared with those of similarly treated normal human leukocytes. Results of the experiment are presented in Figures 2 and 3. Data on Molt-4 cells are presented in Fig. 2. At the end of two hours of exposure, magnetic fields in the presence of holotransferrin significantly decreased the cell count ($p < .014$, compared to control (non-exposed) cells with holotransferrin). Magnetic field alone and holotransferrin alone also showed a slight but significant decrease in cell counts at hour 2. At 22 hrs (hour-24) after exposure, cell counts of control, 'control + transferrin' and magnetic field alone had significantly increased (hour-24 versus hour-2, $p < 0.014$ for all three groups). There was no significant difference among these groups at hour-24. However, cell counts of the 'magnetic field + holotransferrin' treated cells remained low at that time.

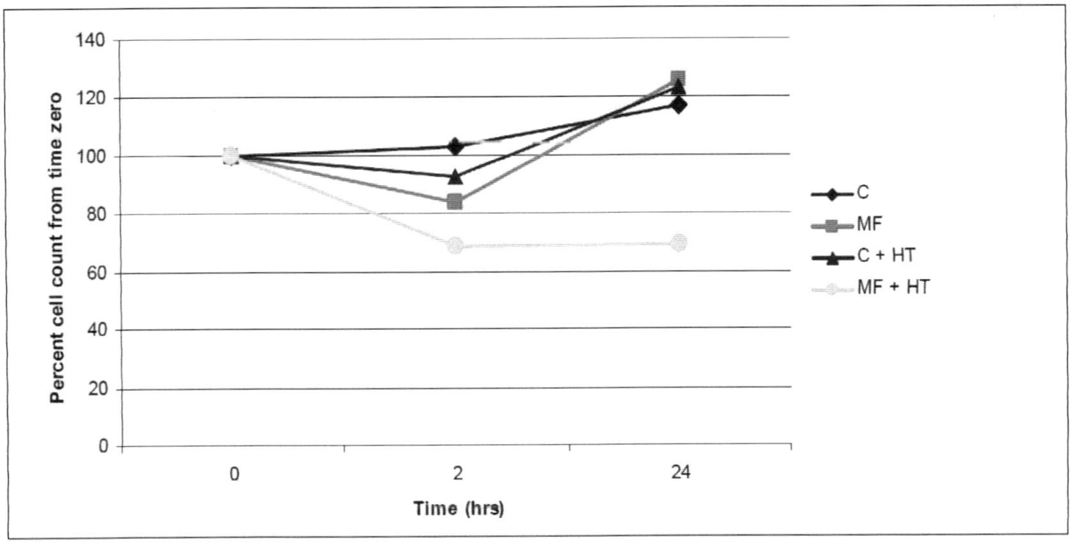

Figure 2. Effects of magnetic field exposure and incubation with holotransferrin on Molt-4 human leukemia cells. Magnetic field exposure started at time zero and lasted for two hours. Each curve represents data from four experiments. C = sham exposed control, MF = magnetic field, HT = holotransferrin.

Data of leukocytes are shown in Fig. 3. At hour-2, there was no significant difference in cell counts among the four treatment groups. Cell counts were significantly decreased at hour-24 (compared to hour-2). This drop in cell count over time is normal for leukocytes in culture. However, the only difference at hour-24 was found between the 'magnetic field + holotransferrin' and 'control' group ($p < .029$). There was no significant difference between 'control + holotransferrin' vs 'control' and 'magnetic field alone' vs 'control'.

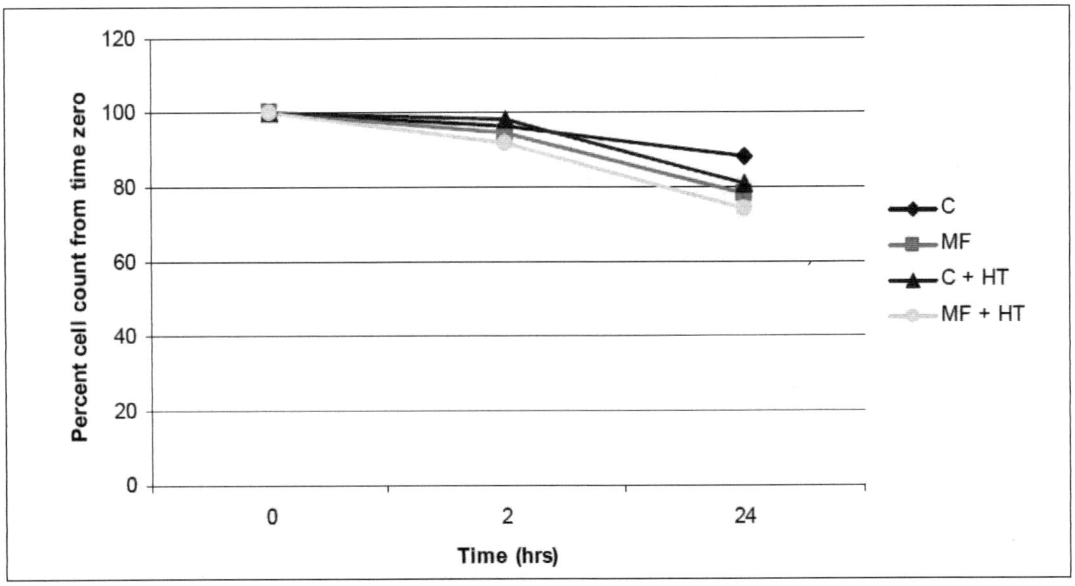

Figure 3. Effects of magnetic field exposure and incubation with holotransferrin on normal human leukocytes. Magnetic field exposure started at time zero and lasted for two hours. Each curve represents data from four experiments. C = sham exposed control, MF = magnetic field, HT = holotransferrin.

Data from this experiment indicate that a 60-Hz magnetic field, in the presence of holotransferrin, has a selective toxic effect on human cancer cells, whereas its effect on normal cells is minimal. This supports our hypothesis that cancer cells are more susceptible to magnetic fields after subjected to a treatment that enhances intracellular iron.

4. Conclusion

The finding that cell counts of Molt-4 cells remained stable at 22 hrs after magnetic field exposure and holotransferrin treatment suggests that the effect of magnetic field is long lasting. The initial decrease in cell count was probably due to cell death as a consequence of DNA and other types of cellular damages triggered by the magnetic field. After that, the remaining cells became stable and stopped proliferating. Free radicals are known to cause cell cycle arrest [22, 23]. A recent study also reports a delay in cell cycling in cells exposed to a 50-Hz magnetic field [24].

Thus, our data indicate that magnetic fields have selective cytotoxic effect on cancer cells, especially under conditions of high iron availability. This can also be applied to in vivo situation. Since transferrin in the circulation is only 30% saturated with iron, it is very easy to increase the availability of iron to cancer cells by simply feeding an animal with a ferrous salt. This will make cancer cells more susceptible to subsequent magnetic field exposure. Oral administration of an iron salt followed by magnetic field exposure can be used for treatment of cancer. Enhanced free radical activity in cells was observed not with 60-Hz magnetic field, similar effects have been reported with

EMF of different frequencies. Thus, EMF of various frequencies could conceivably be used for cancer treatment [25-33].

Thus, EMF may be useful for the treatment of cancer, since cancer cells use high amount of iron and are more susceptible to magnetic fields. Depending on the amount of damage, cell death (apoptosis and necrosis) or cell cycle arrest (e.g., due to DNA strand breaks [34, 35]) can occur. In both cases, the progress of the tumour is retarded.

References

[1] Goldberg D E, Slater A F, Cerami A and Henderson GB 1990, Hemoglobin degradation in the malaria parasite *Plasmodium falciparum*: an ordered process in a unique organelle *Proceedings of the National Academy of Sciences of the United States of America* **87** (8) pp. 2931-2935.
[2] Francis S E, Sullivan D J Jr and Goldberg D E 1997, Hemoglobin metabolism in the malaria parasite *Plasmodium falciparum Annual Review of Microbiology* **51** pp. 97-123.
[3] Yasuhara T, Mori M, Wakamatsu K and Kubo K 1991, Isolation and identification of hemin as an endogenous Na+/K+-ATPase inhibitor from porcine blood cells *Biochemical and Biophysical Research Communications* **178** (1) pp. 95-103.
[4] Tappel A L 1953, The mechanism of the oxidation of unsaturated fatty acids catalyzed by hematin compounds *Archives of Biochemistry and Biophysics* **44** (2) pp. 378–395.
[5] Slater A F G and Cerami A 1992, Inhibition by chloroquine of a novel heme polymerase enzyme activity in malaria trophozoites *Nature* **355** (6356) pp. 167-169.
[6] Slater, A F G, Swiggard W J, Orton, B R, Flitter, W D, Goldberg D E, Cerami A and Henderson G B 1991, An iron-carboxylate bond links the heme units of malaria pigment. *Proceedings of the National Academy of Sciences of the United States of America* **88** (2) pp. 325-329.
[7] Fairlamb A H, Paul F and Warhurst D C 1984, A simple magnetic method for the purification of malarial pigment *Molecular and Biochemical Parasitology* **12** (3):307–312.
[8] Feagin J E, Wurscher M A, Ramon C and Lai H C 1999, Magnetic fields and malaria. (Holick M F and Jung E G Eds). *Biologic Effects of Light: Proceedings of the Biologic Effects of Light Symposium,* Kluwer Academic Publishers, Hingham, MA pp. 343-349.
[9] Wernig F. and Xu Q 2002, Mechanical stress-induced apoptosis in the cardiovascular system *Progress in Biophysics and Molecular Biology*, **78** (2), pp. 105-137.
[10] Lai H and Singh N P 1997, Acute exposure to a 60-Hz magnetic field increases DNA strand breaks in rat brain cells *Bioelectromagnetics* **18** (2) pp. 156-65.
[11] Singh N P and Lai H 1998, 60 Hz magnetic field exposure induces DNA crosslinks in rat brain cells *Singh N, Lai H. 60 Hz magnetic field exposure induces DNA crosslinks in rat brain cells. Mutation Research* **400** (1-2) pp. 313–320.
[12] Lai H and Singh N P 2004, Magnetic field-induced DNA strand breaks in brain cells of the rat *Environmental Health Perspectives* **112** (6) pp. 687- 694.
[13] Lai H and Singh N P 1997, Melatonin and N-tert-butyl-α-phenylnitrone blocked 60-Hz magnetic field-induced DNA single and double strand breaks in rat brain cells *Journal of Pineal Research* **22** (3) pp. 152-162.
[14] Simkó M 2007, Cell type specific redox status is responsible for diverse electromagnetic field effects. *Current Medicinal Chemistry* **14** (10) 1141-1152.
[15] Trowbridge I S, Newman R A, Domingo D L and Savage C 1984, Transferrin receptors: structure and function *Biochemical Pharmacology* **33** pp. 925-932.
[16] Karin M and Mintz B 1981, Receptor-mediated endocytosis of transferrin in developmentally totipotent mouse teratocarcinoma stem cells *The Journal of Biological Chemistry* 256 (7) pp. 3245-3252.
[17] May W S Jr and Cuatrecasas P 1985, Transferrin receptor: its biological significance *Journal of Membrane Biology* **88** *205*–215.
[18] Reizenstein P 1991, Iron, free radicals and cancer, *Medical Oncology & Tumor Pharmacotherapy* **8** (4) pp. 229-233.

[19] Shterman N, Kupfer B and Moroz C 1991, Comparison of transferrin receptors, iron content and isoferritin profile in normal and malignant human breast cell lines *Pathobiology* **59** (1) pp. 19-25.
[20] Castaneda V L, Parmley R T, Saldivar V A, Cheah M S C 1991, Childhood undifferentiated leukemia with early erythroid markers and c-myb duplication *Leukemia* **5** (2) pp. 142-149.
[21] Das-Gupta A, Patil J and Shah V I 1996, Transferrin receptor expression by blast cells in acute lymphoblastic leukemia correlates with white cell count and immunophenotype, *Indian Journal of Medical Research* **104** pp. 226–233.
[22] O'Reilly M A, Staversky R J, Stripp B R and Finkelstein J N 1998, Exposure to hyeroxiz induces p53 expression in mouse lung epithelium *American Journal of Respiratory Cell and Molecular Biology* **18** 43-50.
[23] Powell S N and Abraham E H 1993, The biology of radioresistance: similarities, differences and interactions with drug resistance *Cytotechnology* **12** (1-3) pp. 325-345.
[24] Cridland N A, Haylock R G E and Saunders R D 1999, 50-Hz magnetic field exposure alters onset of s-phase in normal human fibroblasts *Bioelectromagnetics* **20** (7) pp. 446–452.
[25] Hannan C J, Liang Y, Allison J D and Searle J R 1994, In vitro cytotoxicity against human cancer cell lines during pulsed magnetic field exposure *Anticancer Research* **14** pp. 1517–1520.
[26] Hisamitsu T, Narita K, Kasahara T, Seto A, Yu Y and Asano K1997, Induction of apoptosis in human leukemic cells by magnetic fields *Japanese Journal of Physiology,* **47** (3) pp. 307-310.
[27] Miyagi N, Sato K, Rong Y, Yamamura S, Katagiri H, Kobayashi K, Iwata H 2000, Effects of PEMF on a murine osteosarcoma cell line: drug-resistant (P-glycoprotein-positive) and non-resistant cells *Bioelectromagnetics* **21** (2) pp. 112-121.
[28] Ruiz-Gómez M J, de la Peña L, Prieto-Barcia M I, Pastor J M, Gil L and Martínez-Morillo M 2002, Influence of 1 and 25 Hz, 1.5 mT magnetic fields on antitumor drug potency in a human adenocarcinoma cell line *Bioelectromagnetics* **23** (8) pp. 578-585.
[29] Zhang X, Zhang H, Zheng C, Li C, Zhang X and Xiong W 2002, Extremely low frequency (ELF) pulsed-gradient magnetic fields inhibit malignant tumour growth at different biological levels *Cell Biology International* **26** (7) pp. 599-603.
[30] Kirson E D, Gurvich Z, Schneiderman R, Dekel E, Itzhaki A, Wasserman Y, Schatzberger R and Palti Y 2004, Disruption of cancer cell replication by alternating electric fields *Cancer Research* **64** pp. 3288–3295.
[31] Hernández-Bule M L, Trillo M A, Cid M A, Leal J, Ubeda A 2007, In vitro exposure to 0.57-MHz electric currents exerts cytostatic effects in HepG2 human hepatocarcinoma cells *International Journal of Oncology* **30** (3) pp. 583-92.
[32] Kirson E D, Dbalý V, Tovaryš F, Vymazal J, Soustiel J F, Itzhaki A, Mordechovich D, Steinberg-Shapira S, Gurvich Z, Schneiderman R, Wasserman Y, Salzberg M, Ryffel B, Goldsher D, Dekel E and Palti Y 2007, Alternating electric fields arrest cell proliferation in animal tumor models and human brain tumors *Proceedings of the National Academy of Sciences of the United States of America* **104** (24) pp. 10152-10157.
[33] Salzberg M, Kirson E, Palti Y, Rochlitz C 2008, A pilot study with very low-intensity, intermediate-frequency electric fields in patients with locally advanced and/or metastatic solid tumors *Onkologie* **31** (7) pp. 362-365.
[34] Masutani M, Nozaki T, Wakabayashi K and Sugimura T 1995, Role of poly(ADP-ribose)polymerase in cell-cycle checkpoint mechanisms following γ-irradiation *Biochimie* **77** pp. 462–465.
[35] Robles S J and Adami G R 1998, Agents that cause DNA double strand breaks lead to p16INK4a enrichment and the premature senescence of normal fibroblasts *Oncogene* **16** (9) pp. 1113-1123.

IOP Conf. Series: Earth and Environmental Science **10** (2010) 012007 doi:10.1088/1755-1315/10/1/012007

Building Health: The Need for Electromagnetic Hygiene?

Isaac A. Jamieson[1*], Paul Holdstock[2], Helen M. ApSimon[1] and J. Nigel B. Bell[1]

[1]Centre for Environmental Policy, Faculty of Natural Sciences, Imperial College London, London SW7 2AZ, United Kingdom.
[2]Holdstock Technical Services, 3000 Manchester Business Park, Aviator Way, Manchester M22 5TG, United Kingdom.

Email: isaac.jamieson@imperial.ac.uk

Abstract. Whilst the electromagnetic nature of the built environment has changed considerably over the past century, little thought is at present given to the possible advantages of creating electromagnetic microenvironments that more closely resemble those found in nature and/or developing biologically-friendly technology aligned more closely to its operating principles.

This review paper examines how more natural exposures to a variety of electromagnetic phenomena could be re-introduced into the built environment, possible benefits that might arise, and discusses the extent to which there may be tangible benefits obtainable from introducing more rigorous properly considered electromagnetic hygiene measures. Amongst the matters discussed are: the effects of different materials, finishes and electrical items on charge generation (*and the effects of excess charge on contaminant deposition*); the possible benefits of suitably grounding conductive objects (*including humans*) in order to reduce excess charge and contaminant deposition; how the presence of vertical electric field regimes, similar to those found in nature, may enhance biological performance; and possible pitfalls to avoid when seeking to introduce appropriate electromagnetic hygiene regimes.

1. Introduction

This paper investigates the beneficial and detrimental effects that a variety of natural (predominately electromagnetic) phenomena may have on health and wellbeing. It also discusses the extent to which exposure to such phenomena may be altered indoors, and how beneficial natural phenomena (or simulations of them) might, where appropriate, be introduced to improve environmental quality and biological functioning. It is proposed that increased knowledge in this multidisciplinary field may lead to the refinement of the principles under which present day technology and environments are designed and operated.

Several key factors are considered in this work: vertical potential gradients; distorted field regimes; triboelectric charging; charge and particle deposition; grounding regimes; bipolar ionisation; and humidity levels. The effects of different types of natural and artificial lighting regimes are not discussed, though they too can have key roles to play, and will be discussed in detail in a later paper by the lead author.

IOP Conf. Series: Earth and Environmental Science **10** (2010) 012007 doi:10.1088/1755-1315/10/1/012007

2. Vertical Potential Gradients (VPG)

A positive vertical potential gradient (VPG) exists in nature during fair-weather periods, with the positively charged ionosphere acting as an anode and the Earth as a cathode. During such conditions, VPGs of 80-150 V/m can occur in low-lying areas [1], and VPGs of ≤5,000 V/m on high elevations such as on mountains [2]. During 'poor-weather' conditions, triboelectric inversion can occur as a result of air below positively charged areas of clouds becoming negatively charged in relation to the ground directly beneath it causing the direction of current flow to reverse [3, 4].

Some buildings, for instance those predominantly of timber construction, can allow the passage of VPGs into their interiors to a high degree. As an example, traditional rural timber housing can allow ≈70-75% transmission of such fields through their structure [5, 6]. In comparison to this, most modern buildings, particularly those of steel and concrete construction, tend to act as 'Faraday cages' substantially shielding/masking individuals from such fields. The biological effects of shielding from such fields are seldom taken into consideration.

Even centuries old buildings can register zero-fields if electrical conductors and/or large metal uprights are added to them [5]. Indoor environmental factors too can have a role to play. Measurements undertaken by the present lead author indicate that the presence of electrical equipment and materials that gain high charge can also mask such fields (Figure 1).

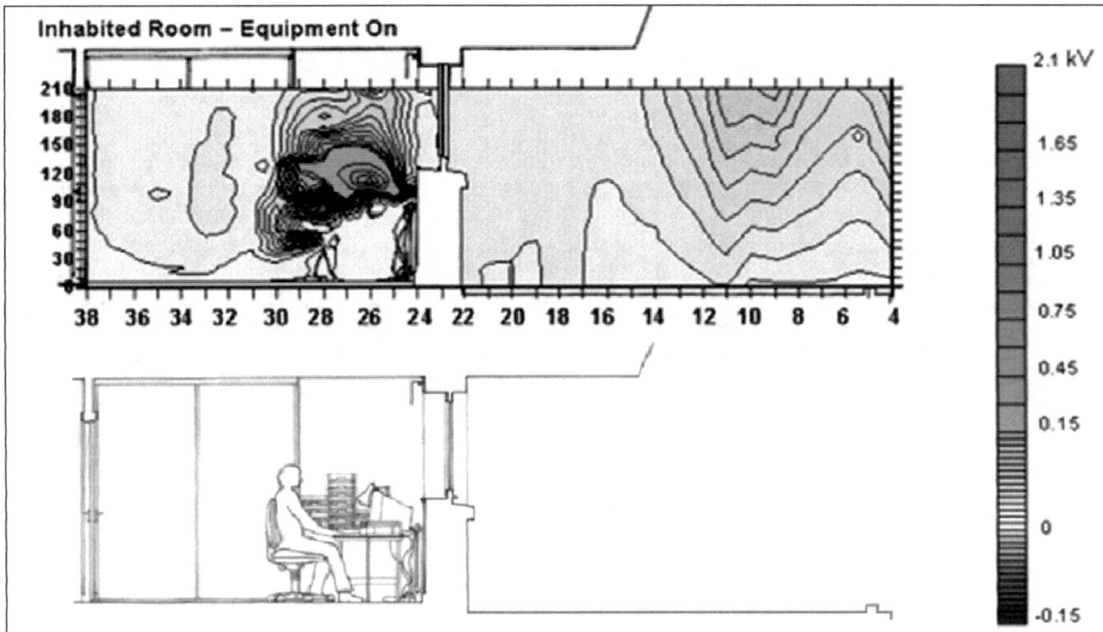

Figure 1. Electrostatic potentials measured indoors and outdoors. (Jamieson – unpublished research).

At present some individuals further shield themselves from VPGs through deliberately creating 'Faraday cage' conditions indoors in an attempt to protect themselves from manmade fields, such as those from mobile phone base stations which can readily penetrate most buildings. Such measures may in part prove counterproductive.

2.1 VPGs and Immune System Functioning

Past research on mice by Möse et al. [7] and Fischer [8] indicate that the screening of VPGs can have pronounced effects on immune system functioning, with exposure to constant vertical DC field protocols significantly altering the operating efficiency of the immune system compared to immune responses provided under Faraday cage and control environments (Table 1).

In that research, the plaque count method devised by Jerne & Norlin [9] was used to determine the degree of immunity exhibited by animals exposed to different field regimes, with higher plaque counts indicating more efficient immune system response. It was revealed that a VPG of only 40 V/m could significantly increase immune system response over true Faraday cage conditions, with further improved immune system responses being noted at the further increased field strengths shown. As mentioned earlier, fair-weather fields of 80-150 V/m are regularly encountered outdoors in nature but are often blocked from ingress indoors.

Number of investigations per exposure	VPG Field strength (V/m)	Plaque counts (indicating degree of immunity)		
		Constant DC field	Control chamber (normal conditions)	Faraday cage conditions
10	40 V/m	210.2 ± 24.1	232.3 ± 13.2	111.6 ± 11.0
10	200 V/m	608.0 ± 55.1	384.0 ± 31.7	199.2 ± 16.5
15	1,000 V/m	572.4 ± 112.8	161.6 ± 31.5	76.0 ± 34.0

Table 1. Plaque Counts after 15-day exposure to various field levels [7, 8].

It is proposed by the present authors that similar improvements in the immune system functioning of humans (and other animals) might occur if exposed to comparable regimes in otherwise low field microenvironments. Whilst improved versions of such regimes might be of particular benefit in hospital environments to improve patient recovery and staff wellbeing, the possibility of introducing them into other building types such as schools, offices and homes should also be considered, especially if electrical items used therein were designed to be more compatible with such systems.

Evidence indirectly supporting possible health benefits that may occur through the adoption of VPG protocols is given by Hahn, who noted dramatic recoveries in sick animals (cows, goat, pigs and a chicken with avian influenza) after exposure to enhanced VPGs. Interestingly, over an 8-year test period, none of the people exposed to those fields (whilst managing the animals on a regular basis) succumbed to colds, influenzas or any another type of infectious disease. Reduced incidents of infection and improved recovery rates were additionally noted in hospital areas with VPG systems installed on their ceilings [10].

2.2 Airborne Contaminants, Microbes and Vertical Electric Field Regimes

The creation of vertical electric fields indoors (*in low field environments* – present authors' emphasis), to replicate those found in nature, has additionally been shown to help reduce concentrations of airborne contaminants and surface deposition in individual microenvironments, and might it is proposed be used as a component in future electromagnetic hygiene initiatives to help substantially reduce individuals' exposures to contaminants whilst increasing general wellbeing.

Past research trials by Hahn [10] demonstrated additionally that VPG regimes could prevent odours and mould growth that had previously occurred in individual offices, and could reduce the airborne concentrations of microbes found in hospital wards by ≈95%. Stersky et al. [11] also demonstrated that VPGs could significantly reduce airborne concentrations of microbes.

The effects of VPG regimes on removing non-biological airborne particulates were investigated by Sandberg & Mellin [12]. Their tests (on a 3.8 m^3 model room) found that the use of a charged ceiling, in situations where there was no ventilation, could reduce concentrations of airborne particulate matter (<250μm) by ≈60% in comparison to control conditions. Additionally, the use of VPGs in conjunction with natural ventilation was shown to reduce particulate concentrations with greater efficiency than that achievable by using natural ventilation alone.

IOP Conf. Series: Earth and Environmental Science **10** (2010) 012007 doi:10.1088/1755-1315/10/1/012007

They also investigated the effectiveness of VPG regimes in removing environmental tobacco smoke (ETS). Typically after < 1 hour, concentrations of ETS were less than they would have been in the control situation where such measures were not been applied. Whilst no quantification was given in their results for that test, such regimes have been used successfully in the past to reduce exposure to ETS in West German restaurants [10], as electric fields are able to act as a major transport and deposition mechanism for submicron particles [13]. A partial listing of contaminants within this size range is shown in Table 2.

Viruses (also arise in larger droplet nuclei)	<0.01–0.31 μm
Diesel soot	0.01–1 μm
Bacteria (also arise in larger droplet nuclei)	0.05–10 μm
Fresh combustion particles	<0.1 μm
Metal fumes	<0.1 μm
Ozone and terpene formed aerosols	<0.1 μm
Environmental tobacco smoke	0.1–0.8 μm
Outdoor fine particles (metals, sulphates)	0.1–2.5 μm
Asbestos	0.25–1 μm
Fungi	0.5–30 μm
Cat dander	1–3 μm
Skin flakes	1–40 μm

Table 2. Common Indoor Particles with Size Range ≤ 1μm diameter [14, 15, 16].

As increased exposures to submicron contaminants are associated with increased prevalence of cardiovascular problems and respiratory diseases, such as asthma and chronic obstructive pulmonary disease (COPD) [17], and may lead to increased likelihood of infection from pathogens, investigation into the possible benefits of applying vertical field protocols in standard indoor situations would appear warranted. It is proposed that the introduction of appropriate VPG regimes indoors could help remove many contaminants from the air whilst improving human performance and wellbeing.

3. Distorted Field Regimes

Even in instances where (natural or artificial) VPGs exist indoors, their presence may be masked in some microenvironments within rooms by highly localised incidences of 'electromagnetic-pollution' as a result of fields created by equipment, finishes and/or materials as indicated in Figure 1.

Research by Jamieson et al [18, 19] suggests that areas which exhibit poor 'electromagnetic hygiene' may often exhibit higher concentrations of charged sub-micron contaminants than low field areas within individual rooms. For the examples shown - of offices measured in Norway and the United Kingdom - a marked reduction in small air ions (SAI) levels occurs in sectors where raised electric fields exist (Figure 2a,b).

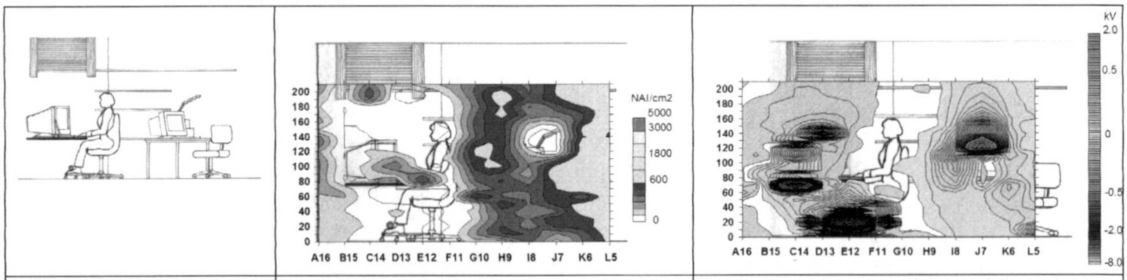

Office	Lower bound of negative small air ion (NAI) levels	Electrostatic potentials in kilovolts (kV)

Figure 2a. Vertical Section through Office Environment. Adapted from [18].

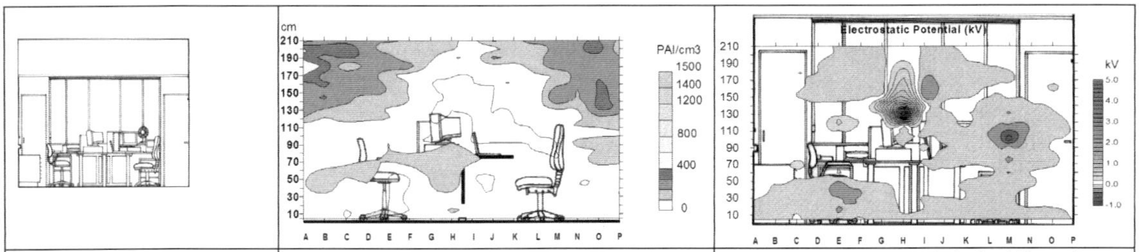

Office	Lower bound of positive small air ion (PAI) levels	Electrostatic potentials in kilovolts (kV)

Figure 2b. Vertical Section through Office Environment. Adapted from [19].

Such conditions are generally indicative of the presence of increased [local] concentrations of harmful charged ultrafine particles (which contribute little to air's conductivity except when SAI are greatly reduced in number). Greater local concentrations of charged neutralised submicron particles that can also cause negative health effects will also occur under such circumstances.

The indication that raised fields, at levels well below those normally considered injurious to health, can increase local concentrations of submicron contaminants (and their localised deposition) appears important, as even small differences in pollutant levels within sections of individual rooms may be vital to medical hygiene and general wellbeing, particularly if exposures to such regimes are prolonged. Often exposures to inappropriate fields regimes (and levels of pollutants) may be easily mitigated.

In addition to optimising the levels and types of fields created, appropriate zoning of field sources may also be undertaken to help improve the electromagnetic character of individual microenvironments [20]. In general, where possible, electrical items that can emit raised fields should be located in areas where individuals are unlikely to spend prolonged periods of time (Figure 3), and/or replaced with suitably designed units that emit low fields. Appropriate finishes and/or materials should also be specified to reduce the likelihood of excess charge.

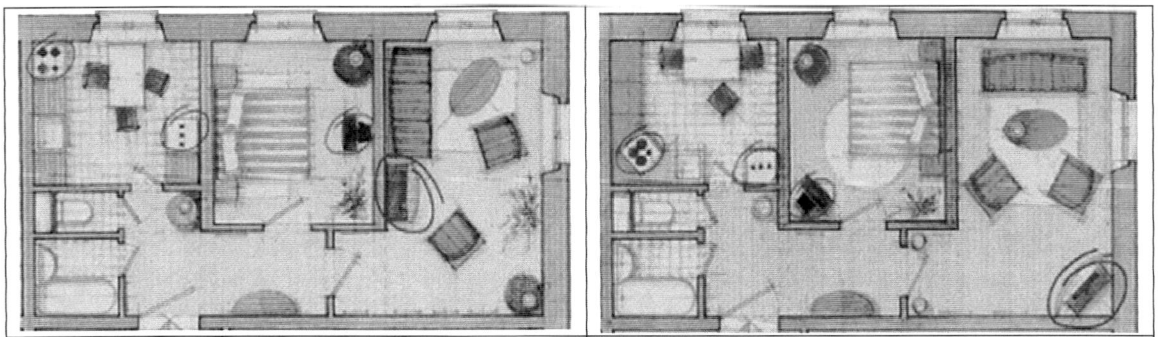

a) Typical room layouts	b) Improved low exposure room layouts

Figure 3. Inappropriate layouts/specification may increase both field and contaminant exposures [20]. (Reproduced with kind permission from the Russian National Committee of Non-Ionising Radiation Protection).

4. Triboelectric Charging of Humans

In the past the human body was often at ground potential for prolonged periods of time. Nowadays, however, high body-potentials can often be readily induced on individuals through exposure to raised potentials from electrical items which have not been designed to take improved electromagnetic hygiene measures into account. Humans and the actions they undertake too are often now a prime cause of excess DC body-potentials, primarily through the triboelectric/contact charging of inappropriately combined/specified modern materials. Tests have shown that the inappropriate use and combination of materials can create charges > 60 kV DC on individuals [21] (Figure 4).

Electrostatic discharge (ESD) incidents can often occur indoors when individuals readily gain potential whilst undertaking many common activities. ESD events can both damage computers and increase local surface contamination and skin contamination rates.

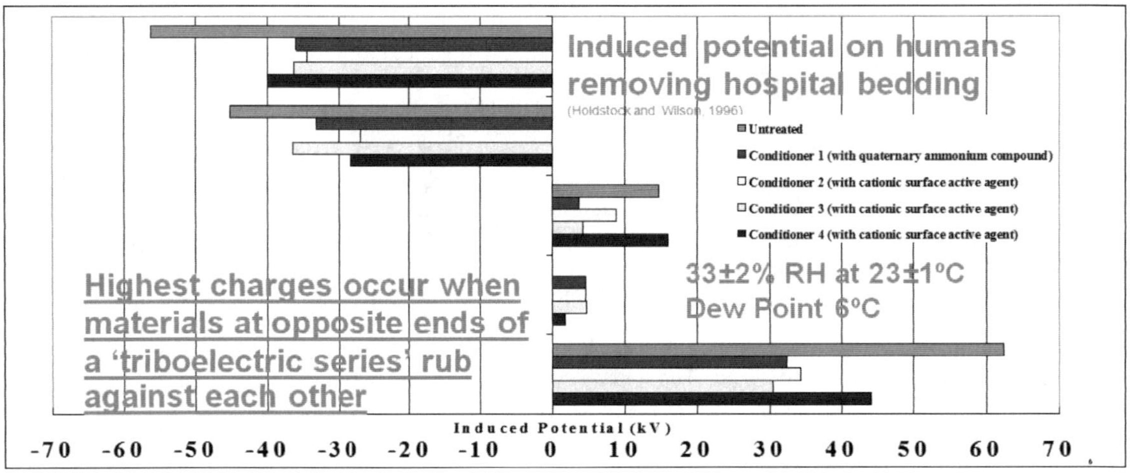

Figure 4. Inappropriate use and combination of materials can create charges >60 kV DC on individuals. (After Holdstock & Wilson [21], with kind permission of the ESD Association).

As the highest incidence of ESD events in hospitals are often found in ward areas [22], and sensitive computer data can be easily lost through ESD events, it appears important as a matter of best practice to ensure that 'mission critical' design measures are followed on hospital wards where computers are used. Another benefit from introducing such measures would be the reduction in localised deposition of submicron contaminants that would otherwise occur through the presence of raised fields.

Allen (23), and Allen et al (24, 25, 26) when conducting research on hospital infections, demonstrated that the size of local electrostatic fields could strongly influence microbe deposition. In one test, deposition was ≤ 20 times greater than control conditions after a 4-day exposure period using voltages ≤ ± 4kV. It was also demonstrated in that work that being in close proximity to highly charged plastic nurses' aprons could greatly increase microbial deposition onto patients.

The raised fields from electrical wiring and equipment (such as televisions and ungrounded laptop computers) that are often found in hospital wards, and elsewhere, should also be taken into consideration when wishing to mitigate against microbial contamination and deposition. In addition to the triboelectric properties of materials and finishes and raised fields from inappropriately designed electrical equipment; poor grounding, humidity levels and air-ion regimes can also contribute to the level of charge created.

5. Influence of Charge on Particle Deposition

As many pathogens, including adenovirus (which causes respiratory infections), influenza A, B and C, SARS-CoV and MRSA [27], are in the submicron range for which electric fields can act as a major transport and deposition mechanism [13], and > 90% of airborne particles in hospitals can be in this size range [28], it appears sensible to discuss the possible influence of charge on particle deposition and how likelihood of infection might be reduced.

IOP Conf. Series: Earth and Environmental Science **10** (2010) 012007 doi:10.1088/1755-1315/10/1/012007

5.1 Infectious droplets and droplet nuclei

Gain of charge, particularly unipolar charge, can directly influence the size of infectious droplets (*such as those containing influenza virons – present authors' comment*), with their disintegration occurring whenever the repellent forces that such charges exert on each other exceed droplet surface tension [29, 30].

The amount of charge held by droplets will greatly determine their behaviour when exposed to electric fields. Whenever the mutual repulsion of charges contained within liquid particles exceeds that of the bonding force of their surface tension, the droplets break-up to form smaller droplets with significantly higher cumulative surface area and overall charging limit. The smaller the particles, the more readily the charge that they hold will influence their deposition. As influenza A tends to infect the lower respiratory system, reduced particle size will aid its deposition in that location.

5.2 Effect of Charge on Respiratory Deposition

Space charge is an important mechanism for increasing respiratory deposition of charged airborne contaminants where high concentrations of charged particles exist. Research by Cohen et al. [31], indicates that singly charged 0.02 μm particles may deposit 5 times more readily than uncharged particles, and 3 times more readily than charge-neutralised particles.

It appears highly likely that areas where raised electric fields occur - such as those shown in [18, 19] - will have higher concentrations of such particles than microenvironments where lower fields are found. It is hypothesised by the present authors that areas with poor electromagnetic hygiene, which can be highly location specific within different microenvironments, can exhibit higher concentrations of charged submicron aerosols and charged-neutralised submicron aerosols thereby increasing the likely deposition rates of such particles in the respiratory tract. Past research has already suggested that improved electromagnetic hygiene in homes may reduce respiratory deposition, with low field environments being shown to help prevent incidents of asthma [32].

5.3 Effect of Charge on Skin Deposition

An indication of how differences in potential can influence the deposition of particulate matter (PM) onto the skin is provided by Wedberg [33, 34, 35] who demonstrated that at body voltages of 0 kV DC, deposition of PM >0.07μm was ≈100 particles/mm^2/hr, compared to ≈1,000 particles/mm^2/hr under moderately high body voltages of ±5-6 kV DC which can be readily achieved in some indoor environments. Deposition rates of >10,000 particles/mm^2/hr were recorded under larger fields.

When it is taken into consideration that excess charge can increase the likelihood of microbial deposition, indications that poor electromagnetic hygiene may greatly increase their localised deposition and deposition onto individuals give cause for concern, particularly when wishing to minimise hospital acquired infections and the spread of infections in other environments. The infective dose required to cause infection varies between microbes; research by Caul [36], for example, showed that the infective dose for Norwalk-like viruses (NLVs) may be ≥ 10–100 infected particles, while Ward et al. [37] noted that for rotavirus (which can cause diarrhoea in young children), the infective dose may be ≤ 10 infected particles.

5.4 Effect of Charge on Surface Deposition

Excess fields increase surface contamination by increasing plate out of submicron contaminants, and by makes them harder to remove, as aerosol deposition velocity increases in proportion to field strength, thereby causing them to exhibit greater deformation and adhesion on impact [38, 39].

Tests have shown that both the particle deposition velocity and the solidity of the materials involved greatly influence final area of the contact site that is achieved by particles, and the strength of adhesive force created. It was noted by Tsai et al. [38], in their research on elastic flattening and particle adhesion, that flattening can significantly enhance adhesive forces, by a factor of ≤ 15 times in soft metals and a factor of > 100 in plastics. The deposition velocity of sub-micron particles, and

IOP Conf. Series: Earth and Environmental Science **10** (2010) 012007 doi:10.1088/1755-1315/10/1/012007

hence their degree of adhesion, is greatly determined by both field-strength and the net-charge that individual particles hold.

As mentioned previously, the research of Allen [23] and Allen et al (24, 25, 26) demonstrates how excess charge can influence microbial deposition.

6. Grounding to Reduce Excess Charge

In most instances local surface-charge should be decreased wherever practicable, as this can reduce the retention and degree of contamination that occurs from sub-micron contaminants in individual microenvironments. In some situations grounding can be used to reduce excess surface-charge, however, the electrical conductivity of the materials being assessed should be taken into account, as electrons can flow freely throughout conductors but not through insulators (Figure 5). Conductors are more likely to gain excess charge through induction whist insulators are most likely to gain it through triboelectric charging and are often a major source of charge generation experienced indoors.

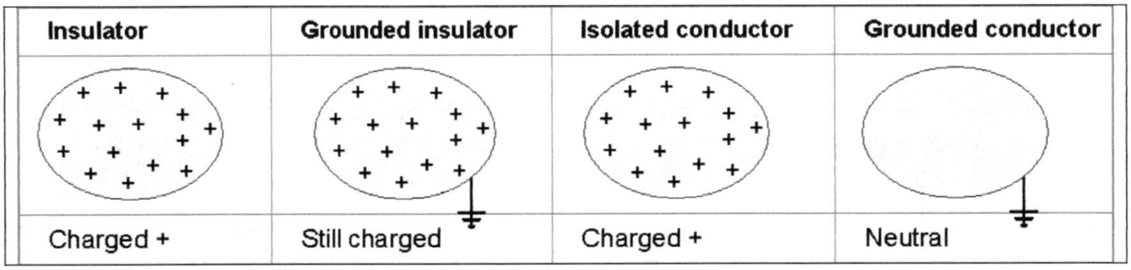

Figure 5. Insulators, conductors & surface charge

Whilst grounding can reduce excess charge on conductive objects, including humans, grounding the human body to reduce excess charge may not always be sufficient to adequately reduce contaminant deposition (including that of microbes) on it when individuals are in microenvironments that have raised electric fields. The potential difference between field sources and individuals is a major factor in determining both deposition-rates and deposition-velocities of sub-micron particulate matter onto the skin. Even when an individual is at zero-potential, a potential difference between the individual and a field-emitter can cause enhanced deposition. The same is true of other conductive objects [40] (Figure 6). If the subject is charged to a potential of opposite polarity to the field source deposition velocities can also be enhanced.

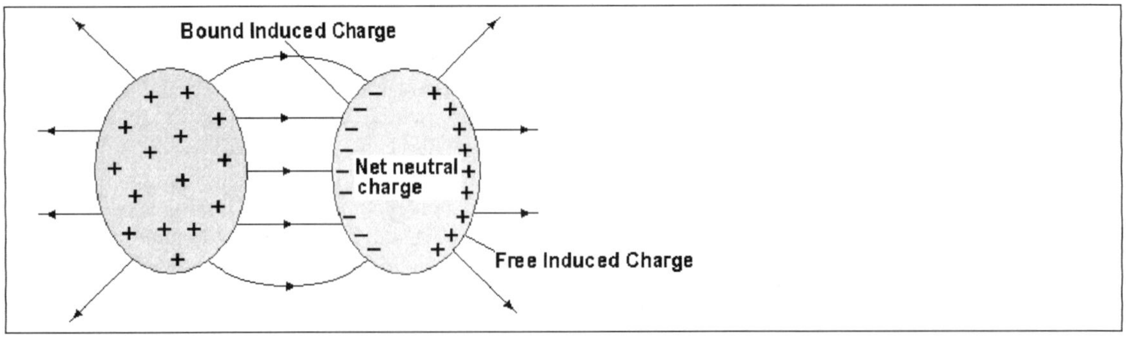

Figure 6. Isolated (ungrounded) conductor exposed to field from charged insulator (After Jonassen [40], with kind permission of Springer Science and Business Media).

Suitably grounding individuals in low field environments can minimise deposition of contaminants onto skin (and it is proposed may also reduce deposition in the airways). In some instances grounding may provide additional health benefits such as stress reduction. As examples of this latter factor, Ghaly &Teplitz [41] found grounding could cause the normalisation of circadian cortisol secretion profiles in some individuals tested, whilst Chevalier & Mori [42] reported that test-subjects appeared less stressed and more attentive when grounded. They also noted improvements in test-subjects' muscle tone when grounded, and that subjects' beta, SMR (slow beta), alpha, theta and delta brainwaves demonstrated abrupt changes after grounding with the right and left hemispheres of the brain coming into synchronicity to a greater degree.

7. Ionisation to Reduce Excess Charge

Bipolar SAI as found in nature can help neutralise excess charge, and the charges on insulators and isolated conductors found indoors can both be neutralised using artificial bipolar ionisation [43] and passive air ionisation measures (such as the specification of appropriate materials that do not gain excess charge readily and the grounding of electrical equipment and conductive items). Even simple "Prudent Avoidance" practices such as occupying lower field microenvironments, and re-zoning electrical equipment within individual rooms, can help improve individuals' exposures to SAI. No direct assessment appears to have been made on how balanced bipolar ionisation - at levels similar to those stipulated in the Russian guidelines for areas where computers are used [44] (Table 3) - may affect the likelihood of deposition of sub-micron contaminants in the airways, or onto the clothing and skin of individuals in comparison to areas with low SAI concentrations.

Minimum	600 NSAI/cm^3	400 PSAI/cm^3
Optimal	3,000 – 5,000 NSAI/cm^3	1,500 – 3,000 PSAI/cm^3
Maximum	50,000 NSAI/cm^3	50,000 PSAI/cm^3

Table 3. Russian Guidelines stipulating mandatory bipolar SAI Exposures for computer workplaces [44].

In addition to being able to reduce excess charge and surface contamination, work by Krueger & Reed [45] has indicated that bipolar ionisation may help reduce the mortality rates of animals infected with influenza (Figure 7). Interestingly, the levels of 2,500-3,500 ions/cm^3 used in that research corresponds closely with the optimum levels stipulated in mandatory Russian guidelines. The levels of <60 NSAI/cm^3 and <100 PSAI/cm^3 used to represent low SAI ion conditions in that research are similar to those sometimes measured at computer workstations in the western world (*present authors' comments*).

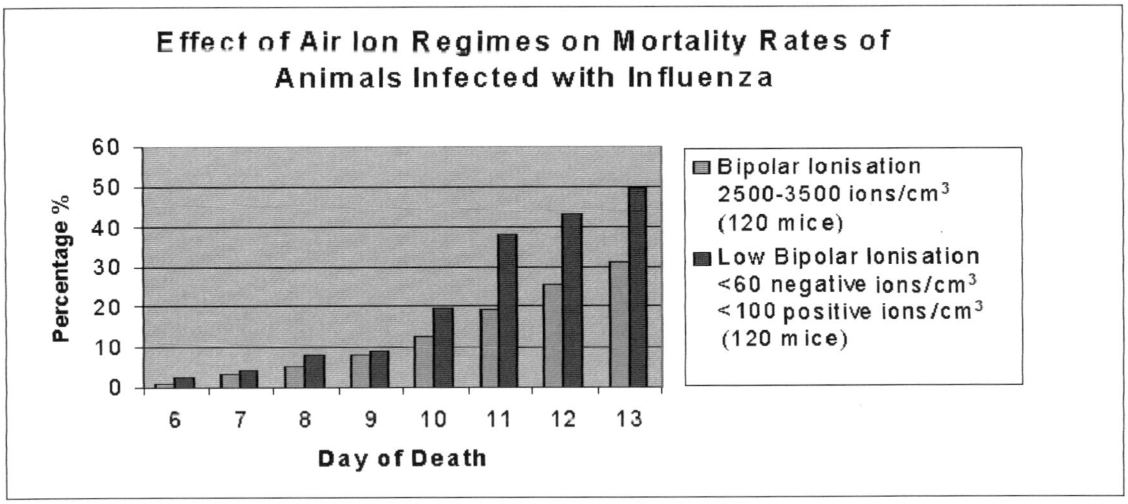

Figure 7. Effect of air ion regimes on mortality rates of animals infected with influenza [45]. (After Krueger & Reed [45], with kind permission of Springer Science and Business Media).

Whilst short-term exposures to standard *unipolar* ionisation have been indicated as being beneficial in some instances, long term exposures to such regimes are not recommended for health reasons [46, 47, 48], with its use actually being prohibited in Russian workplaces [44]. Animal tests in the United States have shown that long-term *unipolar* exposures can result in exposed animals having significantly shorter life-spans than those under non-enhanced conditions [47]. For reasons such as this, SAI regimes more closely resembling those found in nature, such as those recommended in Russia [44], are recommended by the present authors for areas where people spend large periods of time.

8. The Role of Humidity in Electromagnetic Hygiene

Higher levels of charging often occur when the moisture content of the air is <20-30% RH (≈4°C dewpoint temperature), as low humidity reduces surfaces' conductivity (ability to transfer electrical charge), thereby making charge dissipation more difficult [49].

Under such conditions, the higher local electric-field strengths that can be encountered on surfaces that act as insulators or isolated conductors (and the presence of ungrounded electrical equipment) can increase the deposition rate and deposition velocity of many airborne contaminants, including microbes. Low levels of humidity can also make air a component of the charge build-up mechanism as it flows over such surfaces, with dry particles and dust becoming charged to a greater degree on removal from surfaces.

Low humidity levels (though often avoidable with suitable foresight) often occur when air-conditioning is used during summer and winter months, especially in buildings that are not actively humidified [50]. In Scandinavian countries, where low humidity can be a problem (particularly during wintertime), there are often more pronounced cases of screen dermatitis, dry irritated throats and eyes, and cases of electro/chemical-sensitivity due to increased deposition of contaminants onto the eyes, skin and airways of individuals. Reduced humidity can also cause a drying of the skin, making it more conductive [51].

8.1 Humidity and Charge Mitigation

As dewpoint temperature is a more accurate predictor of charge generation than relative humidity (RH) alone [52], it is proposed that dewpoint temperatures should be specified instead of RH, whenever practical, to help mitigate excess charge. Jamieson et al. [19] hypothesise that an 'ideal' average dewpoint temperature for general indoors purposes may be approximately 12°C, and that ≥ 4°C dewpoint temperature might be permissible where this is not possible.

It is further suggested that the application of that proposed 'ideal' dewpoint temperature indoors may help prevent the spread of pathogens such as influenza, particularly during the winter season, as influenza outbreaks normally occur when conditions of low temperature and low RH co-exist [16].

An illustration of the effects that different humidity levels can have on triboelectric charge generation is given by this paper's second author (Figure 8). In this instance, body voltages were measured on an individual after getting up from a desk's main chair, its chair provided for visitors, and walking along a carpet at 11.6°C dewpoint temperature and at ≥ 4°C dewpoint temperature. Marked differences in potential can be seen under the different regimes. As mentioned earlier, increased potential can significantly influence the deposition rates of PM onto the skin [33, 34, 35]. It has also been shown to increase microbial deposition [23]. Optimising humidity levels, along with the other measures discussed, can help to cost effectively reduce such problems.

Low humidity levels can be improved in a number of ways in individual microenvironments, including the use of suitable HVAC systems, commercial humidifiers and/or greenery to help locally regulate the air's moisture content and reduce excess charge. Excessively high humidity levels should be avoided.

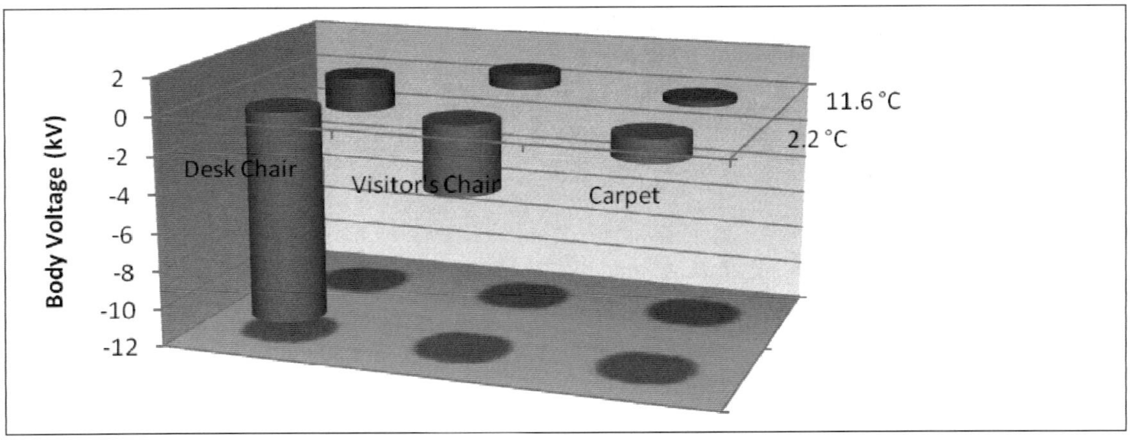

Figure 8. Effects of different dewpoint temperatures on generation of triboelectric charge. (Holdstock – unpublished research).

9. Conclusion

From this review, it appears that there may be a number of substantial benefits achievable for healthcare, industry, general businesses, telecommunications, the general public and other end-users alike through seeking to work with and understand the electromagnetic properties and cues of nature more closely.

The need for further research into electromagnetic hygiene and its related fields appears long overdue, with rapid progress and development of novel (often low cost) technologies and techniques appearing possible in many areas, which could dramatically reduce the socio-economic burden of hospital acquired infections whilst also increasing the wellbeing and work-efficiency of society in general.

Whilst rigorous research is required - and there is a need for caution to ensure that the results of such studies are correctly interpreted and acted upon - the potential rewards for healthcare, businesses, countries, individuals and sustainable technological innovation alike appear great, creating a situation where everyone can gain.

References

[1] Dolezalek H 1985, Remarks on the physics of atmospheric ions (natural and artificial), *International Journal of Biometeorology*. **29**, pp. 211-221.
[2] Jokl M V 1989, Microenvironment, *The Theory and Practice of Indoor Climate*, Thomas, Illinois.
[3] Reiter R 1992, Phenomena in Atmospheric and Environmental Electricity, *Developments in Atmospheric Science, 20*, Elsevier Science Publishers B.V., Amsterdam, 541 pp.
[4] Ogawa T 1995, Lightning Currents, *Handbook of Atmospheric Electrodynamics Volume I*, Volland H (Editor), CRC Press Inc., Boca Raton. pp. 99–136.
[5] Kritzinger H H 1958, Von elektrobiologischen Vorzügen des Holzhauses, *Mitteilungen der Deutschen Gesellschaft für Holzforschung*, Stuttgart, pp. 85 – 100.
[6] Reiter R 1958, Work by Reiter presented by Kritzinger H H 1958, Von elektrobiologischen Vorzügen des Holzhauses, *Mitteilungen der Deutschen Gesellschaft für Holzforschung*, (Stuttgart), pp. 85–100.
[7] Möse J R, Fischer G and Strampfer G 1973, Immunbiologische Reaktion im elektrostatischen Gleichfeld und Faraday-Käfig. *Z. Immunitäts forsch Exp Klin Immunol* **145**, pp. 404-412.

[8] Fischer G 1973, Die bioklimatologisme Bedeutung des elektrostatischen Gleichfeldes. *Zbl. Bakt. Hyg., 1.* Abt., Orig., Reihe B, **157**, pp. 115-130.

[9] Jerne N K and Nordin A A 1963, Plaque formation in agar by single antibody-producing cells. *Science*, **26**, pp. 405.

[10] Hahn F 1956, Luftelektrizität gegen Bakterien für gesundes Raumklima und Wohlbefinden [Air electricity against bacteria for healthy room climate and well-being]. Albrecht Philler Verlag, Minden.

[11] Stersky A, Heldman D R and Hedrick T I 1970, The effect of a bipolar-oriented electrical field on microorganisms in air, *Michigan Agricultural Experiment Station Journal* **5120**, pp. 545-549.

[12] Sandberg M and Mellin A 1987, Artificial electrostatic fields as air cleaning systems in rooms, Proceedings Indoor Air 87, 4th International Conference, Berlin, Federal Republic of Germany, 1987, pp. 231-235.

[13] McMurry P H and Rader D J 1985, Aerosol wall losses in electrically charged chambers. *Aerosol Science and Technology* **4**, pp. 249-268.

[14] Kowalski W J, Bahnfleth W P and Whittam T S 1999, Filtration of airborne microorganisms: modeling and prediction. *ASHRAE Transactions: Research* **105**, pp. 4-17.

[15] McDonald B and Ouyang M 2001, Air Cleaning – Particles. *Indoor Air Quality Handbook*, Spengler J D Samet J M and McCarthy J F (Eds) McGraw-Hill London.

[16] Morawska L 2005, Droplet fate in indoor environments, or can we prevent the spread of infection?, *Proceedings of the 10th International Conference on Indoor Air Quality and Climate (Indoor Air 2005)*, Beijing, China, September 4-9, 2005.

[17] Morawska L, Moore M R and Ristovski Z D 2004, Health impacts of ultrafine particles: desktop literature review and analysis, *Australian Government, Department of the Environment and Heritage, Canberra*, ISBN 0642550557.

[18] Jamieson K S, ApSimon H M, Jamieson S S, Bell J N B and Yost M G 2007, The effects of electric fields on charged molecules and particles in individual microenvironments. *Atmospheric Environment*, **41** (25), pp. 5224-5235.

[19] Jamieson K S, ApSimon H M and J N B Bell 2008, Electrostatics in the environment: how they can affect health and productivity. Electrostatics 2007, *International conference on electrostatics, St Catherine's College, Oxford, UK on 25 - 29 March 2007*, IOP Publishing, Journal of Physics Conference Series 142, 2008, 012052. http://www.iop.org/EJ/article/1742-6596/142/1/012052/jpconf8_142_012052.pdf

[20] RNCNIRP 2007 Russian National Committee of Non-Ionizing Radiation Protection, http://www.pole.com.ru%2fnorm – *in Russian.*

[21] Holdstock P and Wilson N 1996, The effect of static charge generated on hospital bedding, *EOS/ESD Symposium Proceedings 18*. ESD Association, Rome, pp. 356-364.

[22] National Electrical Manufacturers Association in America, Impact of electrostatic discharges in the hospital environment 1995, *NEMA Standards Publication No. SB 29-1994*, National Electrical Manufacturers Association, Washington, USA.

[23] Allen J E 2005, Static electric fields as a mechanism of nosocomial infection. Presentation given at *Electric Fields and Discharges for Microbiology and Healthcare Applications*, conference held by the Electrostatics Group of the Institute of Physics, London, 19 May 2005.

[24] Allen J E, Wynne H, Ross F, Henshaw D L and Oakhill A 2001, Biological aerosols, static charge and hospital infection. In: *Aerosols, Their generation, Behaviour and Applications*, Twelfth Annual Conference of the Aerosol Society, Bath University 18-19 June, 2001. ISBN 0-9529777-6-1, pp. 99-102.

[25] Allen J E, Henshaw D L, Wynne H, Ross F and Oakhill A 2003, Static electric charge may contribute to infections in bone marrow transplant wards, *Journal of Hospital Infection*, **54** (1) pp 80-81.

[26] Allen J E, Close J J and Henshaw D L 2006, Static Electric Fields as a Mediator of Hospital Infection, *Indoor Built Environment*, 15 (1) pp. 49–52.

[27] Kowalski W J, Bahnfleth W P and Whittam T S 1999, Filtration of airborne microorganisms: modeling and prediction, *ASHRAE Transactions: Research*, **105** pp. 4-17. http://www.emergencyfiltration.com/Company.htm).

[28] Rao C Y, Cox-Ganser J M, Chew G L, Doekes G and White S 2005, Use of surrogate markers of biological agents in air and settled dust samples to evaluate a water-damaged hospital. *Indoor Air* **15** (Supplement 9), pp. 89-97.
[29] Wehner A P 1969, Electro-aerosols, air ions and physical medicine, *Special Review in American Journal of Physical Medicine.* **48**, pp. 119-149.
[30] Zhang Y 2005, *Indoor Air Quality Engineering*, CRC Press LLC, Florida, ISBN 1-56670- 674-2.
[31] Cohen B S, Xiong J Q, Fang C-P and Li W 1998, Deposition of charged particles on lung airways *Health Physics* **74**, pp. 554-560.
[32] Bach C 1967, *Ions for breathing: control of the air-electrical climate for health*, Pergamon Press, Oxford.
[33] Wedberg W C 1986, The influence of static electricity on aerosol deposition in indoor environments, aerosols: formation and reactivity, *Proceedings of the Second International Aerosol Conference, West Berlin 22-26 September 1986*, Pergamon Press, Oxford, United Kingdom. pp. 793-796.
[34] Wedberg W C 1987, Facial particle exposure in the VDU environment: the role of static electricity, *Work with Display Units 86,* Knave B and Widebäck P -G (Eds), Elsevier Science Publishers B.V. (North-Holland), pp. 151-159.
[35] Wedberg W C 1991, Risks to VDT operators, Scientific Correspondence, *Nature* **352**, pp. 199.
[36] Caul E O 1994, Small round structured viruses - airborne transmission and hospital control, Lancet, **343**, pp. 1240-1242.
[37] Ward R L, Bernstein D I, Young E C, Sherwood J R, Knowlton, D R and Schiff G M 1986, Human rotavirus studies in volunteers: determination of infectious does and serological response to infection. *Journal of Infectious Diseases* **154**, pp. 871-880.
[38] Tsai C J Pui D Y H and Lui B Y H 1991, Elastic flattening and particle adhesion, *Aerosol Science and Technology*, **15**, pp. 239-255. Cited by Hinds W C 1999.
[39] Hinds W C 1999, *Aerosol technology: properties, behaviour and measurement of airborne particles 2nd ed,* New York: John Wiley & Sons, pp. 260.
[40] Jonassen N 2002, *Electrostatics (Second Edition)*, Kluwer Academic Publishers USA, ISBN 1-4020-7161-2.
[41] Ghaly M and Teplitz D 2004, The biological effects of grounding the human body during sleep, as measured by cortisol levels and subjective reporting of sleep, pain and stress, *The Journal of Alternative and Complimentary Medicine.* **10**, pp. 767-776.
[42] Chevalier G and Mori K 2004, The effect of earthing on human physiology, *California Institute for Human Science,* Unpublished Research Report (completed 03/28/04). Forwarded by C. Ober 2004.
[43] Welker R W 1988, Equivalence between surface contamination rates and class 100 conditions *Proceedings of the 34th Annual Technical Meeting of the Institute of Environmental Sciences*, Mount Prospect, Il, Institute of Environmental Sciences (IES), pp. 449-454.
[44] SanPiN 2003, Ministry of Health of the Russian Federation / Russian Ministry of Health Protection, SanPiN (Sanitary Provisions and Ecological Norms) guidelines 2.2.41294-03 (New SanPiN from 16 June 2003), Air ionization of industrial and public areas according to Sanitary Regulations SanPiN of 22.2.4.1294-03, Appendix 6 - Requirements for air ions content of housings where PCs and VDTs are used – *in Russian.*
[45] Krueger A P and Reed E J 1972, Effect of the air ion environment on influenza in the mouse. *International Journal of Biometeorology*, **16**, (3) pp. 209-232.
[46] Kimura S 1936, Effects of atmospheric electricity on living bodies, *Nippon Onsen Kikogakkai Zasshi* **2**, pp. 108.
[47] Kellogg III E W and Yost M G 1986, The effects of long-term air ion and D.C. electric field exposures on survival characteristics in female NAMRU mice, *Journal of Gerontology*, **41** (2) pp. 147-153.
[48] Baron R A 1987, Effects of negative ions on cognitive performance. *Journal of Applied Psychology*, **72** pp. 131-137.
[49] Vonnegut B 1973, Atmospheric electrostatics. *Electrostatics and Its Applications*, Moore A D (Ed) John Wiley & Sons, London. pp. 390.

[50] Brundrett G W 1977, A review of the factors influencing electrostatic shocks in offices. *Journal of Electrostatics.* **2** 295-315. – Abstract.
[51] Nagda N L and Hodgson M 2001, Low relative humidity and aircraft cabin air quality. *Indoor Air.* **11** pp. 200–214.
[52] Montoya J A 2002, Effect of dew point and relative humidity in electrostatic charge control. *3rd Annual ESD Impact and Control Workshop 14 October 2002,* International Sematech http://www.sematech.org//meetings/past/20021014/agenda.pdf

Disclaimer
This document is intended to help advance knowledge and stimulate further research. It is not intended as a final statement with regard to possible best practice recommendations or potential biological effects. No liability is accepted by the authors for damages arising from its use and interpretation by others. The views expressed do not necessarily represent those of Imperial College London.

IOP Conf. Series: Earth and Environmental Science **10** (2010) 012008 doi:10.1088/1755-1315/10/1/012008

Ways Forward in Public Scientific Controversies

Mike O'Carroll

Vice-Chancellor's Office, University of Sunderland, Chester Road, Sunderland SR1 3SD, UK.

E-mail: mikeocarroll@btinternet.com

Abstract. Public scientific controversies are considered in a defined but fairly general form, essentially where there is uncertainty and divided scientific opinion. General considerations are exemplified in the case of electric and magnetic fields (EMF) from both power supply and mobile phones and other telecoms systems. Characteristics of scientific evaluation and dialogue are discussed. The concept of scientific plausibility is examined and a system of degrees of plausibility is suggested. The paper argues for using graded systems of evaluation in various respects where imposing an inappropriate dichotomy may otherwise make controversies worse. Recognising a plurality of scientifically valid evaluations is recommended and may help to foster public trust in science. Specific steps are suggested for scientists in EMF controversies.

1. Introduction

After more than a decade of involvement in the controversy relating to power-frequency electric and magnetic fields, it is easy to feel frustrated at the lack of progress. It is not for lack of evidence, which has grown steadily, although it is never as conclusive as we might wish. It is not that the arguments are all on the social policy side; at the core they seem to be in evaluation of the science itself.

Science might hopefully be considered convergent, in contrast to politics in an adversarial system. Yet scientists seem to argue just as passionately and without immunity from polarised entrenchment. The result can be that public mistrust of politics and commerce spreads to science.

In a separate contribution to this conference, John Swanson described stakeholder dialogue in the SAGE group (in which I also participated). Stakeholder dialogue embraces both science and social policy. Here I focus more on scientific debate, which SAGE excluded, with a range of other scientific controversies also in mind.

2. Some Basic Definitions

The title calls for defining two things: what is a public scientific controversy and which way is forward?

2.1. A Public Scientific Controversy (PSC)

Define a public scientific controversy (PSC) as a policy dispute involving these parties:

- developers (industry/commerce seeking development, and possibly competitors against);
- public on both sides (e.g. services versus impact);

- scientists on both sides (e.g. on the possibility of health/environment possible risk);
- government deciding on policy (e.g. precaution or exposure restriction).

Controversies where the science is clear and the scientists are all on one side are not included. So a PSC is not about "risk communication" to explain agreed science to an ignorant public. This sort of PSC is defined with EMF in mind but may apply to other public health and environmental issues. The more academic controversies of fundamental physics may escape such a definition.

2.2. PSC State and Ways Forward

Seeking to define ways forward raises ideas of objective, as in strategic planning, and objective-function of state, as in optimisation and equilibrium. How then might the state of a PSC be described?

It is tempting to think of controversy with energy-levels associated with states: from quiet to riot. We're seeking mutual satisfaction, but satisfaction is likely to vary between (and within) parties. The ethics problem of a metric for the common good arises both between and within parties.

In any reasonable ethics metric, we may define a "way forward" as a change which is reasonably expected to improve satisfaction in all parties. Consider PSC state and progress on three fronts:

- towards an objective of an evidence-based decision: clarifying the evidence to support decision options; and agreeing recommendations;
- towards convergence between developer and public: advancing mutual understanding and stakeholder dialogue; mutually acceptable exposure and impact reduction; and mutually acceptable schemes of compensation;
- towards scientific convergence: research, publication and review; improving evaluation and grading systems; and advancing scientific dialogue between opposing scientists.

Here "convergence" is not necessarily to a state of total agreement, but is by degrees to reduce dispersion in the distribution of views.

3. Scientific Dialogue

3.1. Meaningful Dialogue

Define "meaningful dialogue" along the lines of "meaningful consultation": parties listen and give reasons for differences. Unlike consultation, the parties to meaningful dialogue have equal status. In science, the reasons for differences can be analysed scientifically. In spite of the conflicts, science has a culture of convergence when compared with politics.

Overheard in stakeholder dialogue: "it's the scientists that are the worst of the lot" (for being partisan). Early progress may be helped by moderates from both sides.

3.2. Scientifically Valid and Invalid Arguments

Moderates may assist in casting out scientifically invalid arguments on both sides, and in agreeing scientific validity of opposing arguments. Some types of scientific invalidity of *arguments* might include:

- received or religious authority (e.g. creationism, but not necessarily "intelligent design"?);
- being natural or unnatural (e.g. orthomolecular medicine? - but allow evolutionary adaptation to historic exposure);

- romantic ideas (e.g. appeal to tradition, as with astrology or ley lines);
- popularity (appeal to majority views, whether of scientists or of public; e.g. global warming?);
- adverse or favourable consequences of evaluation (e.g. salmonella in eggs, BSE-CJD);
- *ad hominem* criticism (a warning signal of a weak position?).

Note these are scientifically invalid *arguments*. The propositions they support may still be scientifically plausible.

3.3. Some Examples from EMF of the Importance of Definition

In pursuing dialogue it is important to define terms used. Misunderstandings and conflicts can develop from using terms differently.

3.3.1. Information effects. In a Dutch Health Council review 2 September 2008 of the BioInitiative Report, the idea that information, rather than energy, was the key to bio-effects was taken to mean information carried by signal modulation. The Report itself [1] appears to mean information sufficient for bio-detectability of the exposure, which may depend on simple characteristics like frequency rather than energy alone. Neither document defined the term.

3.3.2. Plausibility. Plausibility was famously discussed (with due caution) by Hill [2] as his aspect (6) of association and causation. In a private communication Sir Richard Doll wrote 20 December 1995: "... it is certainly valuable to be made to face the need for clear definitions of the words we use ... Personally I interpret plausible to mean something which I should be prepared to believe. ..."

3.3.3. Degrees of plausibility. In its 2007 review [3] the WHO developed a definition of plausibility at greater length. Section 4.2 of [3] refers to challenge to scientific principles and knowledge, and goes on to refer to signal-to-noise ratios. WHO [3] mentions degrees of plausibility but does not define them.

4. Degrees of Plausibility

In this section, "plausible" is used in the third sense of the Shorter Oxford English Dictionary: having a show of truth ... specious. It is much less than either "credible" or "laudable". In the scientific sense, it will almost mean just being scientifically possible, without prejudice to credibility or truth.

4.1. Plausibility of What?

Ultimately a PSC may be concerned about potential harm, to health or environment. The question of plausibility of harm may depend in turn on plausibility of particular harmful outcomes (like cancer), of possibly harmless intermediate biological effects (e.g. on cell communication), and of specific potential mechanisms for such effects and/or outcomes.

For ultimate harm to be implausible, all potential harmful outcomes should be implausible. Likewise for mechanisms: it is important to allow for possible mechanisms not yet identified. However, when all known potential outcomes, relevant effects and mechanisms have been considered, it can be reasonable to conclude implausibility of harm.

The WHO [3] discusses plausibility essentially of mechanisms, whereas Hill [2] uses the term in a wider sense. Here we consider that it may apply to any hypothesis.

4.2. Features to Consider for Degrees of Plausibility

In building a definition of degrees of plausibility, with harm to health or environment in mind, we might have regard to several features:

- scientific validity: challenges to fundamental scientific principles;
- physical plausibility: signal-to-noise ratio of the external signal;
- bio-physical plausibility: signal-to-noise ratio *in situ* in cells and tissues;
- evidence for lesser or intermediate biological effects;
- evidence for biological capability from contingent areas.

4.3. Scientific Validity and the "Inviolable Laws of Physics"

In defining plausibility the WHO [3] refers to "challenge" to existing scientific principles. A classical example might be the implausibility of a perpetual motion machine. Hill [2] had been more cautious about knowledge of biology while discussing biological plausibility. Knowledge of classical physics may sometimes be regarded as firmer, at least for practical purposes, although bio-physics may present new challenges.

In conference discussion, reference was made to "the inviolable laws of physics" in the context of, for example, mobile phone base stations when the associated public exposure levels seem to be too low. There are dangers in regarding hypothetical effects as implausible on such grounds.

Firstly, the laws of physics continue to be "violated" and re-formed in the face of new evidence. They represent current theory rather than some absolute. However, EMF controversies are not challenging the laws of physics, so that is not an issue in this instance.

Secondly, the idea that exposures might be too low for an effect rests on interpretation of scientific laws, and on understanding potential mechanisms of effects. Appeal to the laws of physics regarding potential effects of EMF may readily apply to non-ionising radiation having insufficient quantum energy to ionise molecules directly; that is not disputed. Rather, one argument is that lower energy excitation may damage biological molecules, including DNA, other than by ionisation. Where the intensity of exposure is low, this properly raises questions of signal-to-noise ratio (SNR).

4.4. Physical Plausibility

Physical plausibility refers to the signal (or exposure) arriving externally to the animal or tissue or person. In air pollution, or in homeopathy, that might refer to concentration. Where this external signal is low (SNR < 1) compared with external noise of the same kind, effects might be considered implausible, or of a very low degree of plausibility.

In environmental power-frequency EMF the external SNR for a signal of 0.4 μT might be of order 1,000, though when compared with typical urban domestic exposure perhaps only of the order of 10. This then drops to 1/10 or less in situ in cells and tissues, due to internal bio-physical noise.

Radio-frequency EMF for mobile phones clearly has physical SNR > 1 in order to secure effective reception. In both forms of environmental EMF exposure there should be no doubt of physical plausibility.

4.5. Bio-Physical Plausibility

Moving from the external signal to a noisier biological environment *in situ* presents a challenge. On the one hand, if a signal can be detected, it can in principle have an effect. On the other hand, the capacity for detection by or in a noisy biological system raises questions of the random nature (or not) and ubiquity (or not) of the identified biological noise, and of biological specialisation and evolution. Such discussions are very active in the field of animal navigation.

It is striking that the SNR for power-frequency EMF reduces sharply from a strong physical SNR externally to well below 1 *in situ* in ordinary tissue. Such bio-physical plausibility is explored by the WHO [3]. Swanson & Kheifets [4] review some 20 EMF mechanisms, concluding: "effects below 5 μT are implausible. At about 50 μT, no specific mechanism has been identified, but the basic problem of implausibility is removed. Above about 500 μT there are established or likely effects from accepted mechanisms".

Yet there are several peer-reviewed reports of biological effects of power-frequency EMF at or below 0.5 μT. In table 2 of section 7 of the BioInitiative Report, Martin Blank lists 8 studies showing several different biological effects at ELF exposures in the range 0.2 – 2 μT. The WHO [3] cites 3 of the 8 references. Swanson & Kheifets [4] cite none of them. In my own casual list of mechanistic results there are a further 6 studies showing effects in practice, or argued theoretically, below 5 μT, only one being cited in [3] and none in [4], though some are too recent to have been cited.

4.6. Evidence for Biological Capability from Contingent Areas

The question of bio-physical plausibility calls for knowledge of biological capability for detecting signals. In turn, biological capability can be informed by contingent study areas. Some examples relevant for EMF signals, yet barely considered, if at all, in EMF reviews, are:

- Health effects of solar and geo-magnetic activity
- Animal navigation
- Magneto-reception in plants
- Cryptochromes
- Berkeley Lab's nano-radio (uses carbon tube of order 100 nm long)

4.7. Tentative Degrees of Plausibility of a Hypothesis (of Harm, or Bio-Effect, or Mechanism)

Since plausibility is a limited component of credibility of a hypothesis, and since the features identified above overlap, a fine and rigid categorisation would not seem appropriate. The following categorisation takes no account of experimental evidence directly supporting the hypothesis in spite of its low plausibility; such evidence would weigh against implausibility when assessing credibility.

4.7.1. Very low plausibility or implausibility. The hypothesis either fundamentally conflicts with accepted scientific principles, or depends upon an external exposure or signal with SNR < 1, or both. There may be exceptions where SNR is approximately 1 where a small increase in noise is likely to bring about the result of the hypothesis.

4.7.2. Low plausibility. The hypothesis is compatible with accepted scientific principles and depends upon an external SNR > 1 with an estimated *in situ* SNR < 1 after taking broad account of relevant biological capability. This category may therefore include bio-physical implausibility. Where knowledge of biological capability is limited, the hypothesis may be classified as of low or reasonable plausibility, depending on the balance of SNR deficit.

4.7.3. Reasonable plausibility. The hypothesis is compatible with accepted scientific principles and depends upon an external SNR > 1 and *in situ* SNR > 1.

5. Degrees of Certainty (or Credibility of a Hypothesis)

In the EMF context the hypothesis is usually of causation of harm. There is a range of uncertainty between rejection and acceptance (or establishment) of a hypothesis. Precautionary action may depend upon the assessment of such uncertainty. It is therefore helpful to have a consistent categorisation of uncertainty and several bodies have attempted this. The question of plausibility (above) is just one component to be considered.

My paper to the Radiation Research Trust conference immediately preceding the IOP conference [5] sets out some "underlying principles for evaluation of science" which address the question of uncertainty. They are not repeated here.

The International Agency for Research on Cancer (IARC) system [6] provides a narrative classification for uncertainty of carcinogenicity, essentially in four categories, but not a probability of causation. It can be very useful to have a weight which can represent degree of uncertainty as a probability. That can then be fed in to risk assessment, as one type of uncertainty (about causation) to combine with the other (about incidence).

The California Department of Health Services [7] deals well with this issue and produces continuous Degrees of Certainty comparable with probability. The Intergovernmental Panel on Climate Change (IPCC) [8], on climate change, likewise does well, and classifies its degrees in a 7-point likelihood scale. Both provide sufficiently fine classifications for practical purposes and give helpful input to the risk analysis stage.

6. Degrees of Everything

Defining terms in a PSC naturally leads to shades of meaning and degrees of strength. Conflicts can develop from dichotomies imposed on graduated variables, particularly if dichotomous judgments are used to exclude valid evidence or viewpoints. This is so both in scientific evaluation and in policy consideration. It can be helpful to use graduated assessments where appropriate. Some further examples, in addition to plausibility and certainty, relevant to scientific analysis are:

- objectivity: recognising that even in experimental observation human interpretation plays a role, all the more so in expert review of a scientific topic, and that there is no unique "weight-of-evidence" method [9];

- quality: recognising that conforming to a protocol or code such as "Good Laboratory Practice" or peer review is neither an infallible nor an exclusive guarantee of quality, and that automatic rejection of "bad studies" by such criteria compromises both evaluation and dialogue;

- independence: recognising that degrees of dependence may attach to authors, laboratories, statutory advisory bodies and scientific reviewers, depending not only on financial interest but also on shared protocols, values and networking (e.g. as discussed in the Hockey Stick Controversy relating to climate change); a fine example in EMF is the relative independence of the freshly appointed Independent Expert Group on Mobile Phones (IEGMP) [10] compared with long-standing internal groups and committees (showing degrees of structural dependence).

7. Ways Forward for EMF Scientists?

This article has focused on scientific debate rather than on social policy issues. It concludes with some steps which EMF scientists might take in order to move forward in the sense defined above. For public scientific controversies (PSCs) more generally, there is much at stake, including public trust in science itself.

7.1. Communication and Dialogue

The usual channels of scientific communication continue to be important and ever more accessible in the internet age. But they will not necessarily provide for meaningful scientific dialogue in a PSC. Conferences such as the present one provide for expression of opposing viewpoints, but not much more. An internet forum can facilitate more prolonged constructive debate; the animal navigation forum hosted by the Royal Institute of Navigation is an example. A longer-term structured dialogue such as the UK SAGE group for power-frequency EMF can be helpful, but would need to include detailed science. Some such structured dialogue would be timely for EMF science for mobile phones and base stations, on an international scale combining internet and face-to-face communication.

7.2. Definitions and Validity

EMF scientists can help by taking care in defining and using terms, and by examining the debate as a whole. In this way they may help to recognise valid opposing arguments and scientifically invalid arguments on their own side. Some areas which might benefit from such an even-handed examination are: bio-physical plausibility; exposures near base stations; irregular "dirty power" exposures; "protective" associations from Interphone studies.

7.3. Graded Evaluation

There are already some grading systems for evaluating uncertain scientific hypotheses. Sufficiently fine grades are needed to relate to risk analysis and to proportionality of precautionary options. It can also be helpful to recognise degrees of other assessments – quality, plausibility etc. – instead of imposing an inappropriate dichotomy.

7.4. Plurality of Valid Scientific Views

At first sight it may seem that a clear agreed unambiguous scientific evaluation in a PSC will be helpful both to policymakers and to public trust in science. However, in areas of uncertainty, which is intrinsic to science, there may be diversity of scientifically valid views. Suppression of valid dissent, however subtle or indirect, is likely to be counter-productive both for public policy and for public trust. Scientists can help by welcoming valid dissent and divergence, and by encouraging policymakers to acknowledge and understand it. Review and advisory bodies should, where appropriate, be able to present plural views with their respective supporting arguments, and policy makers should be able to take account of other reviews and advice in addition to those of their statutory advisors, while giving due weight to each.

References

[1] Blackman C, Blank M, Kundi M, Sage C, Carpenter D, Davanipour Z, Gee D, Hardell L, Johansson O, Lai H, Hansson Mild K, Sobel E, Xu X, Chen G and Sage, S A 2007 *The BioInitiative Report,* http://bioinitiative.org/report/docs/report.pdf

[2] Hill A B 1965 The Environment and Disease: Association or Causation? *Proceedings of the Royal Society of Medicine* **58** pp. 295-300.

[3] WHO 2007 *Extremely Low Frequency Fields (Environmental Health Criteria* **238**) (Geneva: WHO Press).

[4] Swanson J and Kheifets L 2006 Biophysical mechanisms: A Component in the Weight of Evidence for Health Effects of Power-Frequency Electric and Magnetic Fields, *Radiation Research* **165** pp. 470-478.

[5] O'Carroll M 2008 Precautionary approaches and the SAGE experience, RRT Conference: EMF and Health – A Global Issue, London, 8-9 September 2008. http://www.radiationresearch.org/conference/downloads/021520_ocarroll.pdf

[6] IARC 2002 Non-ionizing radiation, Part 1, Static and extremely low-frequency (ELF) electric and magnetic fields (IARC monographs on the evaluation of carcinogenic risks to humans **80**) (Lyon: IARC Press).

[7] CDHS 2002 An evaluation of the possible risks from EMFs from power lines, internal wiring, electrical occupations, and appliances (Oakland: California Department of Health Services)

[8] IPCC 2005 Uncertainty Guidance Note for the Fourth Assessment Report (Geneva: Intergovernmental Panel on Climate Change).

[9] Weed D L 2005 Weight of Evidence: a Review of Concept and Methods *Risk Analysis* **25** pp. 1545-1557.

[10] The Stewart Report 2000 Mobile Phones and Health, *Independent Expert Group on Mobile Phones*, http://www.iegmp.org.uk/report/text.htm

IOP Conf. Series: Earth and Environmental Science **10** (2010) 012009 doi:10.1088/1755-1315/10/1/012009

Intelligent Communication: The Future of EMF Discourse and Risk Governance?

Isaac A. Jamieson

Department of Epidemiology and Biostatistics, School of Public Health, St. Mary's Campus, Imperial College London, London W2 1PG, United Kingdom.

Email: isaac.jamieson@imperial.ac.uk

Abstract. It is proposed that to help facilitate progress in electromagnetic field (EMF) risk governance and actively address potential concerns whilst encouraging environmentally-sound technical advancement, the ground rules of stakeholder participation / debate may benefit from being radically overhauled to encourage more openness, trust, understanding, transparency and innovation.

It is further proposed that two-way communication should be actively promoted, as should proactive thinking and informed discussions of risk (which address realistically the concerns of all parties). The creation/development/adoption of enhanced best practice solutions and precautionary principle measures should also be encouraged.

1. Introduction

There is presently often much discord between members of the scientific community, regulatory bodies, industry and concerned members of the general public over the possible need for more exacting EMF regulations and best practice protective measures.

Some researchers and national bodies (e.g. in China, Italy, Russia and Switzerland) and country regions (i.e. Emilia-Romagna, Toscana and Veneto in Italy) recommend substantially lower EMF exposure limits than are presently suggested in international guidelines [1-6], whilst others are sceptical of the actual benefits of such measures, and often see such proposals as costly, impractical and/or unnecessarily scaremongering. Moreover, whilst some are prepared to consider whether lower exposure levels should be adopted, others wish additional research to be undertaken in advance to try to avoid unsound decisions being made that could have unforeseen negative consequences.

Additionally, whilst in the past little thought has generally been given to the benefits that may arise from scientists and industries working together to create more biologically friendly electromagnetic technologies and environments in the English speaking world [7], such measures are now starting to be seriously investigated again. This paper investigates ways in which EMF stakeholder discourse might be improved to help advance such progress and indicates some of the possible pitfalls to be avoided.

2. Effective Discourse

2.1 Choosing Appropriate Stakeholder Representatives

When the main investigative brief is being set it is important to have the right people in place to determine who should participate and at what level. Ideally, the potential strengths, shortcomings and views of individual stakeholders and those they represent should be factored in so that a level playing field can be set. As noted by Spangler [8], the choice of potential participants is often decided by the size of the undertaking, with many representatives often being predetermined for large-scale matters (as key experts are more likely to attend important national or international meetings than local meetings that affect only a limited number of people).

Robbins [9] suggests that the optimum number of representatives for greatest team effectiveness appears to be between five and twelve, whilst Corder Thompson & Associates [10] states it is more likely to be between eight and twenty. It is often a problem keeping stakeholder groups to a manageable size [11]. For the UK's Stakeholder Advisory Group on ELF EMFs (SAGE) the main group was comprised of 41 stakeholders when its first report was released [12], with six smaller groups also being formed with between 4-15 members participating in each of these groups [6].

Ideally to maximise the chances of meaningful mutually beneficial outcomes (whilst reducing likelihood of complications, conflict and incidences of uncertainty), proper preparation, two-way dialogues and in-depth briefs should be undertaken with appropriately experienced individuals (covering all pertinent viewpoints) as early as possible to try to ensure that proper balanced discussions can take place. Such advanced preparations can help greatly in dealing with any eventualities that may occur as matters progress.

Knowing the subject area and the potential values/costs of decisions to various participants can be of great importance with regard to setting a fair brief, and detecting and preventing biases within the composition of the group at an early stage. Flexibility of opinion may also be a benefit. A variety of skills and attributes should be sought when choosing participants to help proceedings function effectively. These include: technical and analytical ability, agreeableness (pleasant personality and manner towards all participants), people management, openness, problem-solving abilities, ability to listen without filtering information, and conflict resolution skills (where necessary).

It is important to realise from the outset that even single members exhibiting disruptive or irresponsible attitudes towards matters being discussed (or the opinions of other stakeholders) can greatly harm trust, goodwill and potential outcomes as such behaviour can lead to reduced group performance and ability to meet targets [13]. Such states of affair can often occur in instances where conflict is likely to be rife (as is often the case in EMF discourse), indicating the importance of properly vetting stakeholders beforehand. For further details on strategies that can be used to pick appropriate stakeholder representatives refer to Susskind & Cruikshank [14].

2.2 Encouraging 'Active' Listening

Understanding properly what others are trying to fully convey requires constant effort and is often difficult to achieve in practice. Often participants in discussions and stakeholder groups may only 'hear' what others say without 'listening' to it, or seeking to objectively understand that party's point of view and what they are actually trying to convey, thereby greatly prohibiting progress.

'Active' listening, as undertaken by concentrating objectively on what is being said, interpreting the message given, and ensuring that it has been correctly understood (by clarifying with the speaker what they tried to convey) is to be encouraged in all parties and can lead to the creation of goodwill and trust whilst avoiding many pitfalls that might otherwise occur [15, 16]. This courtesy measure is

IOP Conf. Series: Earth and Environmental Science **10** (2010) 012009 doi:10.1088/1755-1315/10/1/012009

important, as when individuals feel that they have not been listened to properly, they are more inclined to disrespect others and distrust what those individuals may say. Also, when time is not taken to adequately address individuals' thoughts and concerns, many opportunities for mutual gain and insight may be lost.

Parties should actively listen to others feelings and needs if they themselves wish to be heard and understood, by becoming an example of how they themselves wish to be treated. Listening impartially to all parties can greatly reduce tension and speed true progress in stakeholder proceedings, especially when parties are able to entertain different opinions and take on board concepts and potential solutions they may not previously have wanted to hear/entertain. As noted by Rosenberg [17], the time necessary to reach conflict resolution can be halved when negotiators agree to paraphrase accurately what other parties have said before replying.

2.3 Attitudes Towards Other Stakeholders

Individuals taking part should seek to treat other parties with tolerance, respect and dignity whilst acknowledging their points of view. If this is not practiced, there is a risk that negative self-fulfilling prophecies may occur that hamper progress and innovation whilst causing unnecessary inter-party conflict and hostility.

Where possible, stakeholders should avoid unreasonable requests and maintain a high degree of transparency between themselves and other parties. Ideally, participants should be encouraged to help remedy possible misconceptions in a non-confrontational or 'constructive' confrontational way - *where individuals argue over concepts and interpretations seeking solutions instead of personalising issues and negatively casting blame and/or making recriminations* - instead of looking down on others who may not be as well briefed, hold alternative views, and/or incorrectly use specialist terminology or concepts. Even experts can be unfamiliar at times with terms and concepts used by other specialists.

Often beneficial insights and goodwill can be gained through introducing gentle corrections instead of dismissing what is being said outright. Nobody is perfect, and many (but not all) are often willing to be corrected if told in an appropriate fashion (particularly if it can be done in such a manner as to prevent them losing face – as can be achieved when individuals try to think of themselves in the other person's position). The attitudes and relationships of the past can be changed.

3. Stakeholder Relationships

3.1 Types of Stakeholder Relationship

Hattori & Lapidus [18], suggest that there are four major kinds of stakeholder relationships that are each dependant to a different degree on trust (Table 1). The first two discussed represent 'Win-Win' approaches that have the potential to be beneficial to all parties, whilst the latter two (commonly used in the past) represent 'Win-Lose' approaches.

Outlook	Relationship	Type of trust	Behaviour	Impetus	Potential outcomes
'Synergy' 'Win-Win' solutions sought for individuals & stakeholder partnership (SHP)	Collaborative	Highly invested openness with good communication	Responsible, with shared end-goals. Seeking to understand others concerns.	Alignment of goals. Mutual concern for success & benefit to all stakeholders	Breakthrough problem resolution – outstanding results & innovation possible
'Win-Win'	Cooperative	Transaction oriented	Willing	Successful SHP outcomes	Preconceived success

'Confrontation' Winning within rules whilst not necessarily being ethical	Competitive	Cautious or hesitant	Shrewd	To look good, meeting own (& not necessarily others) Agenda	Compromise
'Win-Lose' Win at all costs	Adversarial	Distrust	Ruthless	Not to lose	Unpredictable

Table 1. Types of Discourse. Jamieson & Briggs [19] adapted from: Hattori & Lapidus [18].

3.1.1 'Win-Win' Approaches

When wishing to create consensus, it is necessary to consider objectively individual parties' primary objectives, determine what is fair and reasonable in what is being requested by all sides, and then discuss what concessions can be made (if required) as trade-offs by each of the represented interests in order to create what they are hoping to achieve as a collective group.

Dawson [20] suggests that there are four important rules to adhere to when wishing to successfully conduct 'Win-Win' negotiations.

- Avoid narrowing the negotiation down to a single issue, as this creates a 'Win-Lose' situation. As long as more than one issue is on the table trade-offs can occur. Flexibility is important, as is remembering that stakeholders hold shared mutual interests that may be bargained over.

- Stakeholders have different motives, issues and priorities from each other. What is felt important to one stakeholder may not be important to another. Parties should not assume that anything that may benefit another may take away from their own position. It is important, however, to decide what can be conceded to others (without losing ground). Acting in this way encourages others to reciprocate in a like manner.

- Avoid being too avaricious. If negotiations become too unbalanced in favour of one party or set of parties, ill will can be created that can damage the conclusion of the negotiations and the likelihood of future successful negotiations.

- After negotiations are concluded, offer some extra concession that you are not required to give, to create good will.

The 'Win-Win' approaches discussed below encourage participants to reconcile differences within the groups and between each other, contribute towards the discourse, and create pro-active agreements on the best ways forward.

3.1.1.1 'Collaborative' relationships

There are a number of distinct attributes of beneficial trust present in this type of relationship. These include: good communication, genuine commitment to the task in hand; highly invested openness; and ability to deliver what is required. This type of relationship is highly effective in creating innovative solutions, can beneficially revolutionise outcomes for all parties, and can successfully resolve/prevent stakeholders conflicts [21].

Under such discourse there is a mutual concern for the success and benefit of all parties. Shared end-goals exist and the interests of the other participants and the collective stakeholder partnership are considered before those of the party being represented. This type of relationship is not driven by the individual egos of participants, or those they represent, and seeks the optimum solution for all parties taking shared goals into account. As such, it is time consuming and great efforts are made to properly understand the concerns of others and see matters from their perspectives.

The adoption of this type of relationship can lead to rapid and outstanding breakthroughs in problem resolution. Importantly, substantial benefits can be created for all parties through the adoption of this method, or the less labour and time-intensive 'Cooperative' method, in comparison to the more traditional confrontational 'Competitive' and 'Adversarial' methods often adopted in EMF discourse.

3.1.1.2 'Cooperative' relationships
Whilst substantial time and effort are often required to create and maintain effective 'Collaborative' relations, 'Cooperative' relationships are far easier to produce, rely on agreeable behaviour, and may often prove to be more suitable for specific 'one-off' issues. These relationships are again respectful to individual members, and may additionally prove more cost-effective in EMF discourse than the other types of relationships detailed below where lower degrees of trust are found.

'Cooperative' relationships, whilst also encouraging good communication, rely more on a transaction-oriented type of trust than the highly-invested openness found with 'Collaborative' relationships.

3.1.2 'Win-Lose' Approaches
Whilst, in an ideal world, parties might naturally be expected to strive for 'Win-Win' solutions, this often does not arise due to old mindsets and ingrained confrontational ways of thinking which can cause unnecessary and mutually counterproductive conflict to arise. The 'Win-Lose' approach is intrinsically unstable, can cause long-term resentments and misunderstandings, discourages more moderate individuals from participating, and often seeks to benefit (or glorify) one or more parties at the expense of others. The generation of such attitudes and mindsets with these approaches can easily cause counterproductive animosities between those who possess good intentions but opposite perspectives.

3.1.2.1 'Competitive' relationships
The Competitive 'Win-Lose' approach often invokes shrewd behaviour, and a cautious or hesitant form of trust that some perceive as being unethical. Its usage can reduce feelings of goodwill and trust between parties, whilst promoting the likelihood of confrontation and reducing the likelihood of optimum solutions being found.

This form of relationship often results in compromise solutions being created and hampers innovation.

3.1.2.2 'Adversarial' relationships
The Adversarial 'Win-Lose' approach which has often been adopted by many parties in the past involved in EMF discourse can rapidly destroy trust, create 'hardliners', make arguments out of discussions, and cause strong denial as a result of different views and interpretation of scientific research findings. One-sided decisions, or ones which fail to address all the facts available or restrict the brief unnecessarily, can often be counter-productive, cause ill feeling, resentment and distrust.

Hostility, anger, aggression, blame and/or resentfulness can easily occur when individuals feel threatened, or when they feel what they are trying to say is being ignored. Defensiveness can also occur - particularly if individuals are made to feel at fault or in some way lacking - making them more unwilling to listen to others points of view.

When 'hardliners' and other individuals act or react to a situation or person in an aggressive way or make demands (often through anger and frustration, and their perceptions of other people and their objectives) they can inadvertently cause disruption, negativity and ill-feelings in others that further hamper progress and can lead to (often long-term) retaliations for perceived slights. The unscrupulous can create/use such negative situations to create short-term gains for themselves by diverting attention from important matters and/or making others appear unreasonable/reactionary to their colleagues. It is necessary to transcend such patterns.

This 'need to win at all costs' mindset is occasionally pursued by some when their plea for what they believe is appropriate action (or non-action) is ignored/ridiculed or strongly challenged causing them to lose face. In the past it has also (on occasions) been adopted by those in authority when faced with controversial issues or those unwilling to compromise.

Whilst its perceived potential benefits would initially appear limited to only short-term applications, where parties are unlikely to meet each other again, it is often applied successfully – though with diminishing returns - as a mindset for dealing quickly with long-term problems. Confrontational relationships can block much potential progress and the probability of agreement [18]. Additionally, it should be noted that when forced to comply unwillingly with others viewpoints, as opposed to through consensus, resentment can be increased in the losing party creating the likelihood of greater resistance to the others needs in the future [17]. Parties need to accept responsibility for their actions, and ideally seek to undertake ones that will enrich life.

4. Seeking Ways to Improve Stakeholder Partnerships

4.1 An Historical Perspective

As noted by the Stakeholder Advisory Group on ELF EMFs (SAGE) in the UK:

> "*It has been widely acknowledged that public trust and confidence in official information has been an issue ... for some years, not least in scientific advice. There is increasing recognition of the issues concerning biases, conflicts of interest, and independence. ...*
>
> *This widespread lack of trust in scientific advice and allegations of conflict of interest are reflected in the history of EMFs and health. There have been (and continue to be) specific published and unpublished allegations, from both sides of the debate, concerning the validity or integrity of the processes used by review bodies and other bodies making statements about EMFs.*" [6].

As a result of such concerns, and in a wish to certify its impartiality, SAGE sought to ensure that its advisory group would be comprised of relevant stakeholders with interests or relative expertise, primarily on health related EMFs matters, from a wide variety of backgrounds. Among those represented in the SAGE process were: academics, activists, citizen groups, environmental groups, appropriate Government departments (the Department for Communities and Local Government (*formerly the Office of the Deputy Prime Minister*), the Department of Health and the Department of Trade and Industry), Health Protection Agency (HPA) Radiation Protection Division, Industry (the electricity industry, equipment manufacturers, and the mobile phone industry), Office for Gas and Electricity Markets (Ofgem), professional bodies (such as the Council of Mortgage Lenders, the Institution of Engineering and Technology, the Royal Institution of Chartered Surveyors), and research organisations.

Cancer Research UK, housebuilders and developers were asked to become stakeholders but declined. The section of the Department for Communities and Local Government which deals with Building Regulations was also asked to become a stakeholder but did not come on board, though it did contribute to a degree.

The SAGE process itself is financed by the Department of Health, the Electricity Industry and the charity Children with Leukaemia.

4.2 The Need to Ensure Even Handedness and Impartiality

4.2.1 Seeking consensus

Whilst it is normal for stakeholder groups to have a chairperson, a different approach was used within SAGE, with a professional facilitator and assistant being employed instead. Under this holistic approach, which was similar in many ways to the appointment of an ombudsman, encouragement was given to seeking consensus amongst participants.

As part of that procedure a great deal of time was taken to listen and seek to understand the viewpoints of all concerned, allowing all individual matters to be raised and investigated in depth by the stakeholders before decisions were undertaken. Collecting relevant information in that way allowed a sense of shared ownership and trust to be generated. A structured approach has also been employed in SAGE to help identify and evaluate issues raised.

Through the vigorous seeking of consensus, dissenting voices can be heard and not overruled by the majority vote – the voting system does not operate within SAGE. Though it is more time consuming to learn and operate than standard systems (and sometimes less likely to provide quick solutions), listening to dissenting voices can lead participants to the development of improved solutions that otherwise would not have been considered.

However, such processes can be hampered if there are 'hardliners' within stakeholder groups who seek to maintain narrow agendas and adopt rigid positions that can escalate perceived differences. Therefore, as mentioned previously, stakeholders should be picked in advance with great care to help avoid this. Frequently such individuals seek to sabotage attempts by other stakeholders - even those on their own 'side' - to reach compromise and consensus. Many 'hardliners' actually make a living by taking alternative views regardless of the logic or science and can greatly hamper progress. The choice of 'intermediates' and 'moderates' can help address such pitfalls, as they are more likely to perceive valid aspects of the evidence and opinions given by others, and be more open to negotiation on how to best move forward [22, 23].

The use of consensus and the development of common bonds may also help prevent the adoption of short-sighted solutions that might otherwise occur through the adoption of voting measures where majority rule is taken into consideration. In this process individual viewpoints or interests are not allowed to overshadow those of other stakeholders. As an example, for situations where consensus is not realistic, stakeholders in the SAGE process have been able to elucidate on why they disagree, and in so doing circumvent the need for minority reports. Generally, however, considerable convergence has been shown between stakeholders in SAGE, though difficulties were encountered in trying to reach convergence in some areas [12].

4.2.2 Conflict resolution

Conflicts and polarised views can often arise at any stage within negotiations as a result of parties' differing interpretations on evidence given, 'framing' of viewpoints, restrictions placed on matters that may be discussed, or opinions supplied. It is sensible to determine the extent to which these issues have arisen with regard to specific misunderstandings and/or other underlying causes, so appropriate resolution measures can be sought [24].

'Reframing' permits stakeholders with opposing viewpoints to see matters from a different perspective and withdraw from formerly given positions to new shared viewpoints. It can also help participants save face [25]. Whilst in many conflict situations only 'reframing' of participants' standpoints is required to help reach a 'Win-Win' solution, this measure alone may prove insufficient for intractable conflicts. Under such situations, those with opposing viewpoints may be required to examine and better understand the group's underlying needs, in addition to identifying the mutual

IOP Conf. Series: Earth and Environmental Science **10** (2010) 012009 doi:10.1088/1755-1315/10/1/012009

interests of all parties. Parties being amenable to the discussion of formerly prohibited matters can also prove beneficial in helping diffuse conflict.

In cases where agreement still cannot be reached, the opinions of a neutral third party may be sought. If a non-neutral party - *or one that is perceived as non-neutral* - is asked it can destroy trust and goodwill between stakeholders.

Many of the measures outlined above in this paper can also help address these matters before they become a problem, and/or can greatly minimize their potential escalation. For details of further possible intervention measures refer to Burgess & Burgess [26].

5. EMF Stakeholder Group - Proposed 'Best Practice' Measures

As part of the SAGE process, in addition to undertaking cost-benefit analysis, a number of 'best practice' measures and options for reducing exposures to EMFs from wiring in homes, domestic electrical equipment and powerlines were discussed. As it was recognised that there was no single best practice measure to reduce exposures, a variety of possible measures were given. A brief summary of some of those measures, as applicable to dwellings, is given below [6].

5.1 Reduction of Electric Field Exposures

- Dwellings can be wired using cable that has a screen within its overall sheath.
- All concealed wiring can be housed in metal conduits. Where this is not practical, grounded metal screening tape can be applied over cables in ceilings and walls as an alternative. All metal accessories and mounting boxes covers can be used to mitigate fields. String-pull light switches may also be used to help reduce exposures.
- Ideally plug sockets should not be placed in close proximity to beds.
- Demand switches can be installed in order to disconnect the voltage when the circuit is not being used thereby lowering exposures.

5.2 Reduction of Magnetic Field Exposures

- Locate consumer units (and rotating-disc meters) away from areas with high-occupancy. Where possible use electronic electricity meters instead of rotating-disc meters. The meter tails should be contained as physically close together as practical.
- Have radial circuits used for wire power circuits instead of ring circuits.
- Always keep 'go' and 'return' currents physically close to each other. This is especially important for reducing fields from underfloor heating cables and two-way switched lighting.
- Insert sections of plastic piping in runs of metal pipework such as are used for water and gas supplies.
- The whole electrical installation should be protected using a residual current device (RCD).

Additionally as part of their work SAGE recommended that the Institution of Engineering and Technology (IET), which is the successor body to Institution of Electrical Engineers (IEE), should issue guidance to electricians recommending the use of radial circuits as opposed to ring circuits, the benefits of keeping 'go' and 'return' currents physically close together and keeping meter tails together. Refer to SAGE's document on precautionary approaches to ELF EMFs for details of other recommendations made [6].

As noted by Welsh & Welch [27],

"... once a best practice is out there, everybody can imitate it, but companies that win do two things: they imitate and improve."

"[Those] that make best practices a priority are thriving, thirsting, learning organisations. They believe that everyone should always be searching for a better way. [They] are filled with energy and curiosity and a spirit of can-do".

Such mindsets can create large competitive advantages for those who seek to adopt such measures over their rivals.

6. Extending 'Best Practice'

The work undertaken by SAGE, though considered by some to be only marginally successful in terms of creating 'Win-Win' scenarios and taking matters forward, indicates how additional time might be profitably allocated in stakeholder proceedings to further improve stakeholder dialogue and aid the creation/development/adoption of 'graded' best practice and precautionary principle measures related to electromagnetic phenomena in terms of environment, energy efficiency, health and sustainability. Wider brief developments may also be of benefit.

As examples of this concept, as related to the built environment, both the UK Building Research Establishment Environmental Assessment Method (BREEAM) and the internationally recognised US Green Building Council (USGBC) Leadership in Energy and Environmental Design (LEED) Green Building Rating System^{TM}, already assess buildings on a wide variety of environmental and sustainability matters, with credits being given when specific levels of process or performance are obtained.

For BREEAM (and BREEAM International) ratings are given as Pass, Good, Very Good, Excellent or Outstanding, whilst for LEED the ratings are Certified, Silver, Gold or Platinum [28, 29]. The latest version of this is the LEED 2009 rating system - which also takes into account human health concerns - has credits being given for measures that mitigate potential negative health impacts. BREEAM also takes indoor and external issues that may affect health and wellbeing into consideration. Their influence and adoption continues to grow and should also be taken into account in future EMF stakeholder partnerships.

It is suggested that improving stakeholder discourse in such matters may lead to the rapid refinement of the principles under which present day technology is conceived and operated, and buildings and environments are designed. This in turn may create significant technological advancements, product advancements and opportunities for sustainable growth using the 'Win-Win' approach.

7. Conclusions

Coming together for maximum mutual benefit requires strong vision, openness, responsibility, commitment, accountability, fairness, mutual respect, and the wisdom to know how to act appropriately on the findings discovered so that maximum long-term gains are made by all parties.

As indicated above, ideally, the dynamics of EMF stakeholder groups, and framing of agenda, should be reworked in order that parties are able to work together as a true alliance, sharing goals, insights, objectives and a strong desire and motivation to work with each other. Where possible, it is further suggested, that participants should set aside old mindsets and 'destructive' confrontational approaches, and adopt more open proactive non-confrontational and 'constructive' confrontational approaches that sublimate personal goals for the overall benefit of all stakeholders and those they represent.

IOP Conf. Series: Earth and Environmental Science **10** (2010) 012009 doi:10.1088/1755-1315/10/1/012009

Adopting such mindsets has already been shown to create superior results and innovation in some areas of the electronics industry [30], and other situations where conflict has been rife. The 'Win-Win' approach can create significant benefits over the 'Win-Lose' approach presently often adopted in EMF discourse, and may help stimulate progress in technological and sustainability matters at a more rapid pace than would otherwise be achievable.

References

[1] Ahlbom, A, Bergqvist U, Bernhardt J H, Cesarini J, Court L A, Grandolfo M, Hietanen M, McKinlay A F, Repacholi M H, Sliney D H, Stolwijk J A J, Swicord M L, Szabo L D, Taki M, Tenforde T S, Jammet H P and Matthes R 1998, Guidelines for limiting exposure to time-varying electric, magnetic, and electromagnetic fields (up to 300 GHz). International Commission on Non-Ionizing Radiation Protection. *Health Physics*, **74**(4), pp. 494-522.

[2] Chiang H 2003, Rationale for Setting EMF Exposure Standards, Conference Proceedings of the Third International EMF Seminar, held in Guilin, China, October 2003, pp. 134-137.

[3] SanPiN 2003, Ministry of Health of Russian Federation / Russian Ministry of Health Protection, SanPiN (Sanitary and Epidemiological Norms), Computer Workplaces, Sanitary-epidemiological regulations and norms (SanPiNs), National Standards (GOSTs), SanPiN 2.2.4.1191-03, Electromagnetic fields in occupational environment, SanPiN Appendix 3 (Mandatory) – *in Russian*.

[4] Zhaojin C 2006, Comparison of EMF standard and EMF research in China (Abstract), Presentation made at: The Precautionary EMF Approach: Rationale, Legislation and Implementation, convened by the International Commission for Electromagnetic Safety, Benevento, Italy, in February 2006.

[5] Blackman C, Blank M, Kundi M, Sage C, Carpenter D, Davanipour Z, Gee D, Hardell L, Johansson O, Lai H, Hansson Mild K, Sobel E, Xu X, Chen G and Sage S A 2007, The BioInitiative Report, http://bioinitiative.org/report/docs/report.pdf

[6] SAGE 2007, Stakeholder Advisory Group on ELF EMFs: First Interim Assessment, Power Lines and Property, Wiring in Homes, and Electrical Equipment in Homes. Issued 27/04/2007. Accessed at: www.rkpartnership.co.uk/sage/.../SAGE%20supporting%20papers%20first%20interim%20assessment.pdf

[7] Jamieson I A, Holdstock P, ApSimon H M and Bell J N B 2010, Building health: the need for electromagnetic hygiene? *Electromagnetic Phenomena and Health – A Continuing Controversy? – IOP Conference Series: Earth and Environmental Sciences.*

[8] Spangler B 2003, Stakeholder Representatives. *Beyond Intractability*. Burgess G and Burgess H (Eds.), Conflict Research Consortium, University of Colorado, Boulder. http://www.beyondintractability.org/essay/stakeholder/

[9] Robbins S P 2008, The Truth about Managing People (Second Edition). Pearson Prentice Hall Business, London, ISBN 978-0-273-71532-0, 232pp.

[10] CTA 2002, Choosing Stakeholders: A Key Step in a Collaborative Process, Corder Thompson & Associates, http://www.corderthompson.com/documents/Choosing%20Stakeholders%20for%20Public%20Processes.pdf

[11] Gray B 1989, Stakeholder Legitimacy. In: *Collaborating: Finding Common Ground for Multiparty Problems*. Jossey Bass Business and Management Series, Jossey Bass Publishers, San Francisco. pp. 66-69.

[12] O'Carroll M 2008, Precautionary approaches and the SAGE experience, *RRT Conference: EMF and Health – A Global Issue*, London, 8-9 September 2008.

[13] Felps W, Mitchell T R and Byington E 2006, How, when and why bad apples spoil the barrel: negative group members and dysfunctional groups. *Research in Organizational Behaviour*, **27**, pp. 175-22.

[14] Susskind L and Cruikshank J 1987, Breaking the Impasse: Consensual Approaches to Resolving Public Disputes. Basic Books, New York, 288pp.

[15] Robbins S P and Hunsaker P L 2006, Training in Interpersonal Skills: TIPS for Managing People at Work (Fourth Edition). Prentice Hall, ISBN-10: 0-13-148151-7, 416pp.
[16] Robbins S P 2008, The Truth about Managing People (Second Edition). Pearson Prentice Hall Business, London, ISBN 978-0-273-71532-0, 232pp.
[17] Rosenberg M B 2003, Nonviolent Communication: A Language for Life (2nd Edition), PuddleDancer Press, United States of America, ISBN 978-1-892005-03-8, 222pp.
[18] Hattori R A and Lapidus T 2004, Collaboration, trust and innovative change, *Journal of Change Management*, **4** (2), pp. 97–104.
[19] Jamieson I A and Briggs D J 2009, Towards Effective Risk Discourse: the Role of Stakeholder Partnerships, *International Journal of Risk Assessment and Management*, **13** (3-4), pp. 276-293.
[20] Dawson R 1999, Secrets of Power Negotiating: Inside Secrets from a Master Negotiator (Second Edition), Career Press, Franklin Lakes, New Jersey.
[21] Foxworthy R 1997, Trust Based Leadership TM, Workbook and training materials, Foxworthy Consulting, Inc., Haymarket, VA – cited by Hattori & Lapidus 2004.
[22] Maiese M and Burgess H 2003, Extremists and Spoilers, *Beyond Intractability*. Burgess G and Burgess H (Eds), Conflict Research Consortium, University of Colorado, Boulder. http://www.beyondintractability.org/essay/extremists/
[23] Maiese M 2004, Moderates, *Beyond Intractability*, Burgess G and Burgess H (Eds). Conflict Research Consortium, University of Colorado, Boulder. http://www.beyondintractability.org/essay/moderates/
[24] Moore C W 1996, The Mediation Process: Practical Strategies for Resolving Conflict (Second Edition). Jossey-Bass Conflict Resolution Series, Jossey-Bass Publishers, San Francisco.
[25] Rosenberg S 2004, Face, *Beyond Intractability*. Burgess G and Burgess H (Eds), Conflict Research Consortium, University of Colorado, Boulder. http://www.beyondintractability.org/essay/face/
[26] Burgess G and Burgess H 2003-2006, Beyond Intractability Version II, The Beyond Intractability Knowledge Base Project, Conflict Research Consortium, University of Colorado, Campus Box 580, Boulder, CO 80309. http://crinfo.beyondintractability.org/checklists/public-policy_intermediaries.jsp?nid=5084
[27] Welch J and Welch S 2005, Winning, Harper Collins Publishers, London, pp 184-185.
[28] BREAM 2009, BREEAM: the Environmental Assessment Method for Buildings around the World, The (UK) Building Research Establishment http://www.breeam.org/
[29] LEED 2009, LEED 2009: Technical advancements to the LEED rating system, U.S. Green Building Council, http://www.usgbc.org/DisplayPage.aspx?CMSPageID=1971
[30] Hiltzik M A 1999, Dealers of Lightning: Xerox PARC and the Dawn of the Computer Age. Harper Business, HarperCollins Publishers, New York.

Acknowledgements

In addition to the work carried out by SAGE, many of the matters and concepts discussed in this paper are drawn on the concepts, theories, opinions and ideas of participants in the joint workshop on risk assessment and risk communication, organised by the SEGH (Society for Environmental Geochemistry and Health) and the EU-funded INTARESE (Integrated Assessment of Health Risks from Environmental Stressors in Europe), HEIMTSA (Health and Environment Integrated Methodology and Toolbox for Scenario Assessment), 2-FUN (Full-chain and Uncertainty Approaches for Assessing Health Risks in Future Environmental Scenarios) and NoMIRACLE (Novel Methods for Integrated Risk Assessment of Cumulative Stressors in Europe) projects, in Athens in April 2008.

Many of the matters and concepts discussed are also drawn on the concepts, theories, opinions and ideas of participants at the conference 'EMF & Health - A Global Issue: exploring appropriate precautionary approaches' organised by the Radiation Research Trust, and 'Electromagnetic

IOP Conf. Series: Earth and Environmental Science **10** (2010) 012009 doi:10.1088/1755-1315/10/1/012009

Phenomena and Health - a Continuing Controversy?' held by the Electrostatics Group of the Institute of Physics. Both of these events were held in London in September 2008.

Grateful thanks are given to all those participants, and to Professor David Briggs and Dr Kees De Hoogh at the Department of Epidemiology and Biostatistics, Imperial College London, United Kingdom. In particular, special thanks are given to Professor Mike Repacholi at the Department of Electronic Engineering, University of Rome 'La Sapienza', Rome, Italy for his helpful comments and suggestions during the preparation of this paper. I also acknowledge support from the EU 6th FP funded project INTARESE for part of the work presented here.

Institute of Physics
Dirac House, Temple Back
Bristol BS1 6BE UK

CURRAN ASSOCIATES INC
proceedings.com